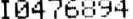

LARGE-SCALE RENEWABLE ENERGY GUIDE

Developing Renewable Energy Projects Larger Than 10 MWs at Federal Facilities

A Practical Guide to Getting Large-Scale Renewable Energy Projects Financed with Private Capital

Cover photos, clockwise from the top:

Installing mirrored parabolic trough collectors – (January 19, 2012) Crews work around the clock installing mirrored parabolic trough collectors, built on site, that will cover 3 square miles at Abengoa's Solana Plant. Solana a 280 megawatt utility scale solar power plant (CSP) under construction in Gila Bend, Arizona, USA. When finished it will generate 280 MW 's of clean, sustainable power serving over 70,000 Arizona homes. *Photo by Dennis Schroeder, NREL 20097*

Dry Lake Wind Power Project in Arizona; Suzlon S88 wind turbines – The 63-MW Dry Lake Wind Power Project in Arizona is the first utility-scale power project. The Salt River Project is purchasing 100% of the power from the Phase I of this project for the next 20 years. The project is located just southwest of I-40 and Holbrook. *Photo courtesy Iberdrola Renewables, Inc., NREL 16701*

Aerial view of the 2 MW PV system at U.S. Army Fort Carson – Aerial view of the 2 MW PV system at U.S. Army Fort Carson financed through a Power Purchase Agreement (PPA). *Photo Courtesy U.S. Army Fort Carson, NREL 17394*

McNeil Generating Station of Burlington Electric Department – The ever increasing demand for electrical power leads public utilities, such as the Burlington Electric Department (BED), to find new, more efficient ways to generate electricity. Wood gasification has the potential to meet this requirement by converting the wood into a gaseous, energy intensive fuel source that can be used in high efficiency gas turbines to generate electric power at low cost. A proposal currently under discussion for a planned gasification project at Burlington Electric's McNeil station in Burlington, VT, has as its overall objective the demonstration, at large scale, of Battelle's High-Throughput Biomass Gasification Process as an efficient, low cost means of generating electric power from wood. Shown here, wood chips are stockpiled in the foreground and the gasifier building is pictured in the background. *Photo by Warren Gretz, NREL 04745*

Solar Two proves the technology is in place for producing utility-scale power from the sun when you need it – during periods of peak electricity demand by consumers – Supplying 10 MW—enough to power 10,000 homes—to Southern California Edison Company's electric distribution grid during periods of peak demand, Solar Two is proving the value and technical capability of power towers. *Photo by Warren Gretz, NREL 02157*

Executive Summary

To accomplish Federal goals for renewable energy, sustainability, and energy security, large-scale renewable energy projects must be developed and constructed on Federal sites at a significant scale with significant private investment. For the purposes of this Guide, large-scale Federal renewable energy projects are defined as renewable energy facilities larger than 10 megawatts (MW) that are sited on Federal property and lands and typically financed and owned by third parties.[1] The U.S. Department of Energy's Federal Energy Management Program (FEMP) helps Federal agencies meet these goals and assists agency personnel navigate the complexities of developing such projects and attract the necessary private capital to complete them.

This Guide is intended to provide a general resource that will begin to develop the Federal employee's awareness and understanding of the project developer's operating environment and the private sector's awareness and understanding of the Federal environment. Because the vast majority of the investment that is required to meet the goals for large-scale renewable energy projects will come from the private sector, this Guide has been organized to match Federal processes with typical phases of commercial project development. FEMP collaborated with the National Renewable Energy Laboratory (NREL) and professional project developers on this Guide to ensure that Federal projects have key elements recognizable to private sector developers and investors.

The main purpose of this Guide is to provide a project development framework to allow the Federal Government, private developers, and investors to work in a coordinated fashion on large-scale renewable energy projects. The framework includes key elements that describe a successful, financially attractive large-scale renewable energy project.

This framework begins the translation between the Federal and private sector operating environments. When viewing the overall effort of both parties in this framework, four key points are clear:

1. The efforts of Federal agencies, private developers, and financiers are inter-dependent.

2. Federal agencies can play a large role in reducing project risk and thereby attract developers and private capital investment.

3. Each party's operating context, constraints, and language must be acknowledged by the other.

4. Successful projects are often the result of each party working together to define a common goal and an understanding of each other's terminology and processes.

Defining Success: A Common Goal

The Federal Government and the private sector renewable energy developer share a common goal: to deploy significant amounts of large-scale renewable energy projects on Federal lands using private capital financing. Federal statutes and Executive Orders have set forth requirements and goals for renewable energy use by the Federal Government; the scale of this effort is very large. The Department of Defense (DOD) has set a goal of deploying 3 gigawatts of renewable energy on Army, Navy, and Air Force installations by 2025. The Army prepared a solicitation estimated at $7 billion, to buy renewable energy from projects financed by the private sector.[2] Thus, private financing must be available to achieve these goals. Meanwhile, developers and investors demand a return on their investments. Renewable energy projects have proven to be profitable, so investors, eager to find new markets, will be interested in the opportunity that large-scale renewable energy projects on Federal agency lands present.

A Common Language

Establishing a working relationship between Federal agencies and private developers is complicated by the fact that the language of each is very different, even unrecognizable, from the other. Behind this language barrier, however, both parties have processes and procedures designed to produce measurable results, limit wasteful effort or spending, and provide transparency to those investing in the effort. These underpin the common goals and intentions while the similarities of purpose drive the parties to overcome the differences and forge a common language that creates the effective, essential communication for a successful working relationship.

---

[1] These projects may include utility-scale facilities, which connect to the electric transmission system and have a primary consumer (off-taker) besides a Federal agency as well as commercial-scale facilities that may be interconnected to the grid but have the Federal agency as the primary off-taker.
[2] http://www.usace.army.mil/Media/NewsArchive/StoryArticleView/tabid/232/Article/3057/army-announces-7b-multiple-award-task-order-contract-request-for-proposal.aspx

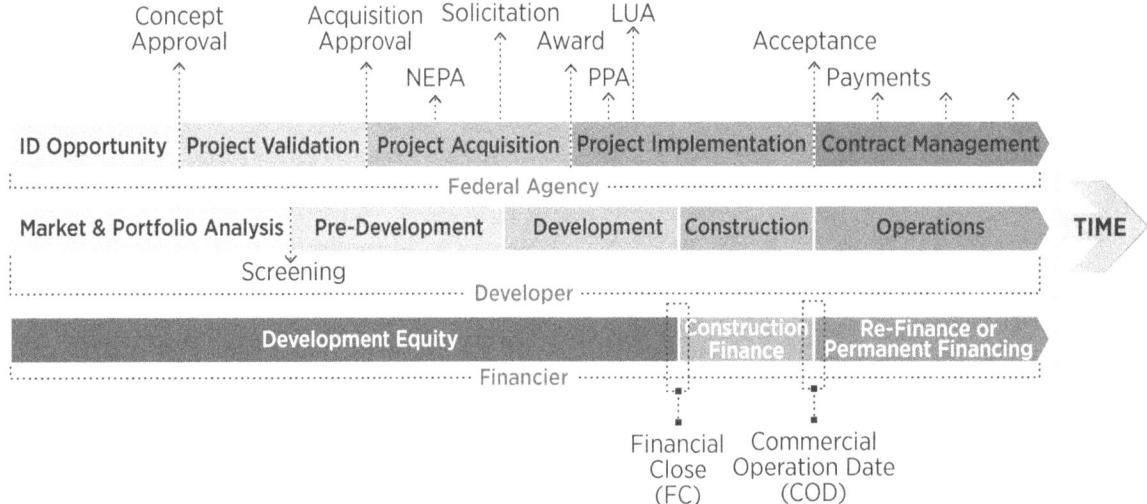

Figure 1. Developing a common language [3]

Figure 1 provides an overall view of some of the similarities of process and differences in language from three key perspectives: that of the Federal agency, the private developer, and the financier. This translation between the three key parties involved in procuring and supplying privately financed renewable energy projects is the starting point for the development of effective communication and a successful project.

A Common Process

To achieve the shared goal of large-scale renewable energy production, Federal agencies and private developers must recognize that they are working as separate parts of a larger, common project development and finance process. Like members of a relay team, the two parties must be synchronized at the point of exchange to be successful, requiring a common framework that recognizes the overall project development environment while accommodating the unique requirements and constraints of each party's operating context. The developer and financier assess and manage risk throughout the project; Federal agencies must do the same.

Under certain methods of financing, the party assuming the lead role of the project will change throughout the life of the project. Accordingly, to be successful, both parties must share a common view of the project's viability. For Federal agencies and employees, this means making early-stage decisions that make projects attractive to the private financial markets and minimizing both real and perceived project risks when possible, so that projects are competitive when taken to the market for development and project financing. For private developers, it means understanding the intricacies of Federal requirements and financing options and being prepared for the Federal process to run its course.

As developers and Federal agencies begin to recognize their contributions as part of a larger, continuous process, continuity in approach and methodology will begin to emerge. A framework to visualize a common process is shown in Figure 2, which is discussed in further detail in Section II (A Reliable, Repeatable Project Development Process).

The framework in Figure 2 was established by NREL and is based on widely used commercial practice. Federal agencies can benefit from understanding the principles of this framework; some may recognize certain elements manifesting in their current practices. The framework becomes a system made up of the following four elements:

1. A timeline of the project development lifecycle that maps influential forces (risk and unknowns).

2. A framework that introduces seven categories of project development, used to organize information.

3. A process that runs across the framework, analyzing project potential (iterative, fatal flaw).

4. Tools used to support decision making, including financial pro formas (Appendix F. Project Pro Forma Example) and development checklists (see Appendix B. Project Development Framework Categories).

Note that it is not possible to show all the detail in Figure 2. Key project milestones such as site control (e.g., land use agreement [LUA]), off-take contract (e.g., power purchase agreement [PPA]), Large-Generator Interconnection Agreement, and Permits will occur on a sliding scale during the "Development" part of the process, depending on the characteristics of the Developer and its corporate objectives. In addition, the National Environmental Policy Act (NEPA) activity shown is if the project is well defined with regards to location and size. If the project is to be defined by the project developer, the NEPA process would start after the solicitation/selection of the project and be completed prior to

---

[3]Acronyms defined: NEPA = National Environmental Policy Act, PPA = Power Purchase Agreement, LUA = Land Use Agreement.

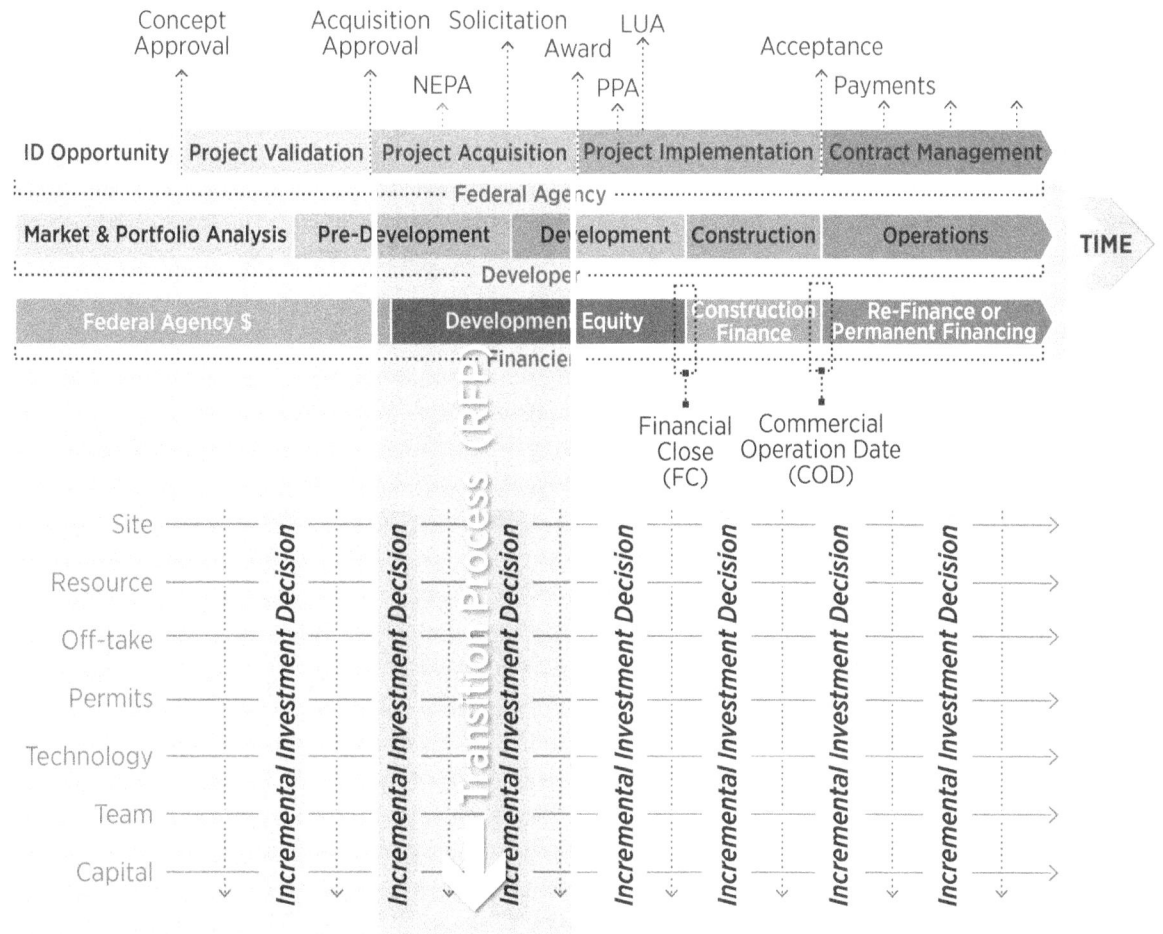

Figure 2. Developing a common process [4]

final agency project approval. The commercial project financing of a large-scale project would typically not reach "Financial Close" without these project attributes in place; however, a company can independently decide the amount of risk it is willing to take during the development process in allocating development capital to achieve project milestones, meet contractual obligations, and respond to changes in market conditions (such as pricing or competitive bidding). See "Limitations of Figures" in Section I, for additional discussion of the simplification in this graphic.

The success of Federal renewable energy projects depends on the ability of both parties to recognize each other as essential to reaching a common goal. Neither party will be successful if the requirements and constraints of the other are not met and understood. The methodologies of each party must be merged or

translated so that a common language, purpose, and process are developed and maintained between all parties.

This Guide acts as a first step in facilitating the process of financing certain types of large-scale renewable energy projects by beginning to translate the differences in language and by mapping a process that is grounded in the foundations of commercial project development while integrating traditional Federal methodologies. An organizational framework, evaluation process, and sample tools are provided for Federal employees seeking to benefit from or gain insight into private development methodologies.

FEMP intends to update this Guide regularly to improve Federal capabilities in this nascent field.

---

[4] Figure 2 provides a general description of each party's responsibilities throughout the stages of project development. However, each party's responsibilities will depend on the type of financing employed (e.g., the private sector party has the lead role in the project development stage in the Energy Savings Performance Contract context). Please refer to the FEMP website for guidance on specific types of financing: http://www1.eere.energy.gov/femp/financing/mechanisms.html.

This Guide is not legally binding and it does not provide legal advice or explanations. Anyone with additional questions should seek appropriate and qualified counsel.

Please check the FEMP website for the latest developments at *www.femp.energy.gov.*

Acknowledgements

The Federal Energy Management Program (FEMP) would like to acknowledge the contribution and assistance provided by the staff of the U.S. Army Energy Initiatives Task Force (EITF). This framework was developed in concert with the EITF's process for large-scale renewable energy project development. FEMP, NREL, and the EITF have worked collaboratively and in parallel on these processes.

FEMP also thanks the agencies that provided input on early drafts and those who provided comment to the Federal Register draft for their time and thoughtfulness.

Table of Contents

Executive Summary 1

    Defining Success: A Common Goal 1

    A Common Language 1

    A Common Process 2

Acknowledgements 4

Table of Contents 5

List of Figures and Tables 8

Introduction 9

    Purpose 9

    Background 9

    How This Guide is Set Up 10

I. Language 11

    Alignment and Sequencing 12

    Limitations of Figures 12

II. A Reliable, Repeatable Project Development Process 13

    Project Development Stages 13

        Stage 1. Market and Portfolio Analysis 14

        Stage 2. Pre-Development 16

        Stage 3. Development 16

    Applying the Framework in a Project Environment (Considering the Financier's View Point) 18

        Risk in the Development Equity Phase 18

        Iterative Approach/Incremental Investment 20

III. Application of Project Development by a Federal Agency 22

    Making Federal Projects Attractive to the Private Sector 22

    The Federal Process 23

        Stage 1. Identify Opportunities 24

        Stage 2. Project Validation 25

        Stage 3. Project Acquisition 26

        Stage 4. Project Implementation 27

        Stage 5. Contract Management 27

    Conclusion 27

IV. Outlook 29

V. Points of Contact 29

Additional Resources...29

Appendix A. Portfolio Approach...30

Portfolio Then Project...30

Portfolio Analysis Steps..30

Strategy..31

Competition and Discipline...31

    Competition..31

    Discipline..31

Appendix B. Project Development Framework Categories...32

B1. Site...32

    Pre-Development Stage Site Elements..32

    Development Stage Site Elements...32

    Inter-relationships: How Site Issues Affect Other Project Development Elements................32

    Example Project Questionnaire – For Federal Sites...33

B2. Resource...35

    Pre-Development Stage Resource Elements...35

    Development Stage Resource Elements..35

    Inter-relationships: How Resource Issues Affect Other Project Development Elements........35

    Example Project Questionnaire - Resource...36

B3. Off-take..36

    Pre-Development Stage Off-take Elements...37

    Development Stage Off-take Elements..37

    Inter-relationships: How Off-take Issues Affect Other Project Development Elements.........37

    Example Project Questionnaire - Off-take..38

B4. Permits including NEPA Compliance and Permitting Activities...39

    Pre-Development Stage Permit Elements...39

    Development Stage Permit Elements..40

    Inter-relationships: How Permit Issues Affect Other Project Development Elements...........40

    Example Project Questionnaire - Permits...40

B5. Technology..41

    Pre-Development Stage Technology Elements..41

    Development Stage Technology Elements...41

    Inter-relationships: How Technology Issues Affect Other Project Development Elements....41

    Example Project Questionnaire - Technology..41

B6. Team 42

    Pre-Development Stage Team Elements 42

    Development Stage Team Elements 42

    Inter-relationships: How Team Issues Affect Other Project Development Elements 42

    Example Project Questionnaire - Team 43

B7. Capital 43

    Pre-Development Stage Capital Elements 43

    Development Stage Capital Elements 43

    Inter-relationships: How Capital Issues Affect Other Project Development Elements 43

    Example Project Questionnaire - Capital 43

Appendix C. Overview of Electricity Markets and Key Terms 45

    Utility Market Structures 45

    Balancing Authority Areas 45

    Transmission and System Operators 46

    Congestion 46

    Ancillary Services 46

    Renewable Markets and RECs 46

    National Organizations 46

Appendix D. Commercial Project Financing 47

    What Makes a Deal? 47

Appendix E. 10-Step Project Development Framework Approach 49

    Coordination of Government Phases and 10-Step Process 50

        Stage 1. ID Opportunity 50

        Stage 2. Project Validation 50

        Stage 3. Project Acquisition 50

        Stage 4. Project implementation 50

Appendix F. Project Pro Forma Example 52

Appendix G. Project Risk Assessment Template 67

Appendix H. Project Validation Report (DRAFT) 69

Appendix I. Summary of Responses to Comments. (DRAFT) 71

List of Figures and Tables

Figure 1. Developing a common language 2

Figure 2. Developing a common process 3

Figure 3. The language barrier 11

Figure 4. Developing a common language 12

Figure 5. Project lifecycle phases 13

Figure 6. Project development stages, translated between government and private sectors 14

Figure 7. A financier's perspective of project lifetime and milestones 18

Figure 8. Process translation adding the financier's language 19

Figure 9. Project phases with risks and unknowns 19

Figure 10. A single iteration in a project development framework 20

Figure 11. Iterations and incremental investment decisions lead to financial close 21

Figure 12. Project stages 23

Figure 13. A typical government process for certain types of financing 24

Figure 14. Transition between Federal sector role and private sector role across project lifecycle 28

Figure E-1. Project development framework with 10-step process 49

Table 1. Financial Model Basic Elements 52

Figure G-1. Example Army project risk assessment framework 67

Figure G-2. Example Army project risk assessment template 68

Introduction

Purpose

This Guide has been created to help Federal agencies effectively develop large-scale renewable energy projects[5] at Federal facilities. For the purposes of this Guide, large-scale Federal renewable energy projects are defined as renewable energy facilities larger than 10 megawatts (MW) that are sited on Federal facilities, property, and lands, and are typically financed and owned by third parties. Because these projects often rely on private investment, it is necessary for Federal agencies to understand the types of large-scale renewable energy projects that the private sector is pursuing. In other words, if the projects that need private sector funding do not attract the private sector, they will never be built. Therefore, this Guide provides the Federal employee with an understanding of a common process that private sector developers use to select projects for investment.

Federal agencies and the private sector share a similar overall process but use and understand different languages. This language barrier masks the similarities in each party's overall process; different language can prevent Federal employees from understanding the important details in the developer's process and vice-versa.

This Guide, while written primarily for the Federal employee, will also be relevant to private sector renewable energy developers and financiers interested in participating in the Federal market. Its scope is limited to large-scale renewable energy project development on Federal lands or facilities in which the energy is consumed by or sold to the Federal facility, a utility, or another project participant.

This Guide is distinct from many other documents on Federal energy projects by de-emphasizing the contracting methods used to execute the project. The purpose of this Guide is to describe the fundamentals of a successful, financially attractive, large-scale renewable energy project. If a project is solid, it is likely that one of several contracting mechanisms can be used to execute a deal. Projects may be funded by private financing through one or more of the project funding options available to the Federal sector,[6] including Power Purchase Agreements (PPA), Energy Savings Performance Contracts, Utility Energy Service Contracts, Enhanced Use Leases, and others. Project procurement would occur through some form of competitive offer framework, often using a solicitation or Request for Proposals (RFP) format or other appropriate mechanism.

This Guide is not legally binding and it does not provide legal advice or explanations. Anyone with additional questions should seek appropriate and qualified counsel.

Background

The United States Government is committed to increasing its consumption of renewable energy and allowing more renewable resources to supply the utility grid. This requires the development of large-scale renewable energy projects at Federal facilities. By deploying large-scale renewables, the Federal Government contributes to energy independence and security, environmental protection, and economic development.

Renewable Energy Use Requirements, Goals, and Related Guidance

1. Beginning in FY 2013, the Federal Government has a goal for not less than 7.5% of the total amount of electric energy consumed to come from renewable energy.

   - 42 U.S.C. § 15852(a) (EPAct 2005, section 203).
   - http://www1.eere.energy.gov/femp/regulations/epact2005.html#rer.

2. Each agency shall ensure that at least half of its renewable energy consumption comes from "new" renewable sources (placed into service after January 1, 1999) and to the extent feasible, the agency implements renewable energy generation projects on agency property for agency.

   - Executive Order 13423, section 2(b).
   - http://www1.eere.energy.gov/femp/regulations/eo13423.html

3. DOE Federal Energy Management Program Renewable Energy Requirement Guidance for EPAct 2005 and E.O. 13423.

   - www.eere.energy.gov/femp/pdfs/epact05_fedrenewenergyguid.pdf.

4. Each agency shall increase agency use of renewable energy, implement renewable energy generation projects on agency property, and prepare targets for agency-wide reductions of greenhouse gas (GHG) emissions. (Federal renewable energy projects implemented on-site may contribute to each agency's scope 1 and scope 2 GHG reduction targets.)

   - Executive Order 13514, sections 2(a)(ii), 7(b)(i).

5. Individual agencies may also have agency-specific goals. For example, DOD has a 25% goal beginning in 2025.

---

[5] These projects may include utility-scale facilities that connect to the electric transmission system and have a primary consumer (off-taker) besides a Federal agency, as well as commercial-scale facilities that may be interconnected to the grid but have the Federal agency as the primary off-taker.

[6] This Guide describes a general process on how to develop large-scale renewable energy projects at Federal facilities using private capital. It does not, however, discuss the approaches that are to be followed under any specific type of financing. For more information on financing a large-scale renewable energy project using a power purchase agreement, energy savings performance contract, utility energy service contract, enhanced use lease, or other method of financing, please refer to guidance on the FEMP website as well as other Federal regulatory materials. http://www1.eere.energy.gov/femp/financing/power_purchase_agreements.html

Federal energy policies, requirements, and goals involve levels of renewable energy consumption that are estimated to require the development of as much as $20 billion of renewable power projects over the next decade. Federal law authorizes, and the current administration has emphasized, the use of private capital to make these investments. These investments will be re-paid under the various types of financing methods available to the Federal Government. These include long-term PPAs, or other energy services agreements, whereby the government, a utility provider, or other project participant will purchase the energy produced by the projects installed and operated on Federal lands.

The definition of renewable energy for Federal facilities is based on the language in the legislative renewable goal for Federal agencies in EPAct 2005 section 203 (42 U.S.C. § 15852(a)). As described in detail in FEMP's guidance for that goal, Renewable Energy Requirement Guidance for EPAct 2005 and Executive Order 13423, Federal renewable energy includes electric energy generated from solar, wind, biomass, landfill gas, ocean, geothermal, waste to energy, new incremental hydroelectric generation at existing plants, or hydrokinetic energy. In addition, the Department of Defense (DOD) has a different renewable goal of 25% of facility energy use by 2025 (10 U.S.C. § 2911(e)), last amended by the National Defense Authorization Act (NDAA of FY2012) with slightly different definitions allowing thermal energy and energy from ground source heat pumps, but not including hydrokinetic energy.

This guide is focused on renewable energy generated to provide energy to a Federal facility or utility. It is not a guide for developing biofuel plants to produce fuel for vehicles. The intent of this document is to present general steps that should be applicable to any project that would generate energy at a Federal facility. In addition to providing energy, renewable energy projects can help agencies reduce greenhouse gas emissions under Executive Order (EO) 13514. The Federal definitions for renewable energy only change slightly for hydropower under EO 13514.

While the renewable energy industry has experienced rapid growth around the world, the industry and its business models may still be unfamiliar to the capital markets. This tends to limit the pool of investors that are willing to participate in renewable energy projects, as does the tax driven nature of the investment requirements. Global economic concerns significantly limit the types of risks and projects that investors will consider, constraining the ability of renewable energy project developers to attract financing for new market opportunities.

These market conditions present both a challenge and an opportunity to the U.S. Federal sector as it strives to attract closely held project development and project finance funds from the private sector. Banks, which are essential participants in the project capital markets, often talk about the "flight to quality" when referring to an investor's appetite for projects, a phrase referring to a very strong preference for low-risk deals.

A long-term agreement with the Federal Government can be seen as an attractive, low-risk revenue stream for the developer that may garner a corresponding low cost of financing for the

developer and a resulting lower energy cost for the government. This may spur more Federal renewable energy projects because power prices will generally be lower with lower costs of financing, while a well-developed market will influence competitive behavior, driving innovation and keeping prices down.

The challenge for developers is that unfamiliarity with the Federal contracting process can result in the perception of tremendous development risks. Given the condition of financial markets, developers are likely to choose transaction partners with whom the development risk is perceived to be lowest. To flourish, the Federal sector should be seen as a viable market segment for project development investment funds.

To efficiently attract private capital, projects on Federal property should be well defined with manageable and financeable development risk that is consistent with market conditions.

Large-scale renewable energy projects on Federal lands should be competitive with other project investment alternatives and attract a broad range of investors. Interest from the financial community to provide long term financing in turn attracts developers willing and ready to put private development capital at risk. The commitment of developers and investors together significantly increases the likelihood of project completion and overall quality of the end result.

> To efficiently attract private capital, projects on Federal property should be well defined with manageable and financeable development risk that is consistent with market conditions.

How This Guide is Set Up

Section I (Language) of this document highlights the language barrier between Federal agencies and the private sector— this language translation continues throughout the document. The section is not meant to be comprehensive; instead, the intent is to begin the process of developing a common language between parties. Once the Federal employee understands some of the private sector terminology, he or she can start to better understand the developer's process and how it relates to the government's process.

Section II (A Reliable, Repeatable Project Development Process) describes a process commonly used by the private sector to develop large-scale renewable energy projects.

In Section III (Application of Project Development by a Federal Agency), the government employee is provided with a reasonable understanding of what his or her responsibility and/or role is within the context of the large-scale renewable energy project development process, while attempting to provide the developer community with a recognizable, reliable, and predictable process in which it can engage with a reasonable likelihood of commercial success. Each of the appendices provides more detail on the subjects covered in sections two and three.

I. Language

The intent of this Guide is to minimize the language barrier that may exist between Federal agencies and private developers, and to highlight the importance of government agencies and private developers' understanding of each other as they navigate the project development and execution process. Language provides a foundation for understanding and trust between the parties involved in project development. It can also be a barrier to success if disparate jargon is used without thoughtful translation. Because the Federal agency is, in all likelihood, contracting with a project developer, and not directly with the financier, this Guide will focus on the language of government and the private sector developer.

Each party's language is formed and developed with respect to the unique perspective and context of the party involved, both of which are driven by different (but compatible) motivations and constraints. The essential activity between the parties is to come together in a mutually beneficial relationship that achieves a common goal that neither party can obtain alone. In other words, the goal is to get a deal done.

Beyond the elements of scope, schedule, quality, and budget (fundamental concepts of project management), the development and procurement of a renewable energy project introduces both risk and financial issues that are complex and lasting. These issues cannot be negotiated efficiently while language barriers exist. An illustration of the differences in context and languages between Federal and private parties is shown in Figure 3, which shows both parties joined by a common goal (deploying large-scale renewable energy projects), but potentially undermined in achieving that goal by the uncertainty created by different perspectives and language.

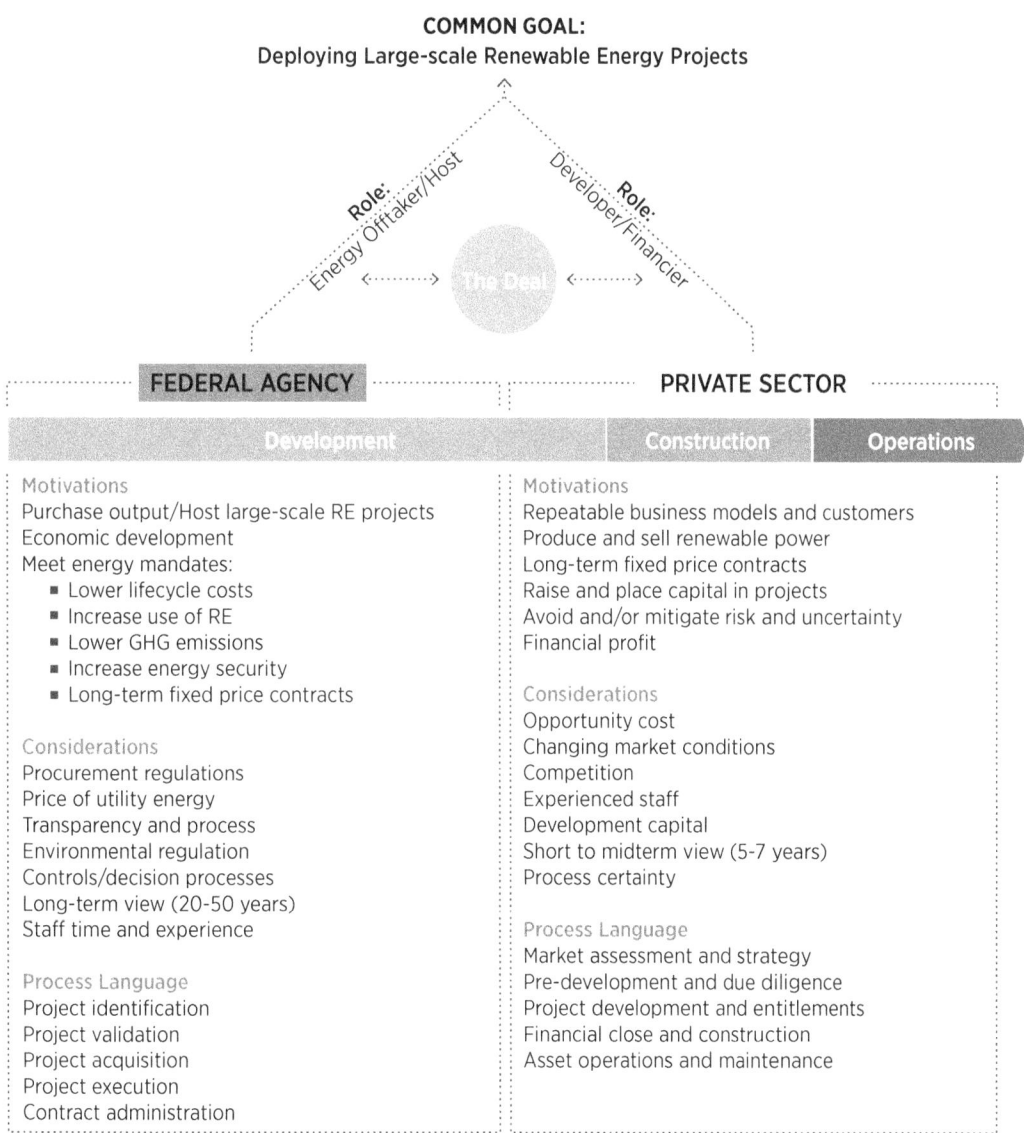

Figure 3. The language barrier

Behind these differences, however, both parties have processes and procedures designed to produce measurable results, limit wasteful effort or spending, and provide transparency to those investing in the effort. These are the underpinnings of common goals and intentions.

Figure 4 shows an overall view of some of the potential similarities of process and differences in language between the Federal agency and the private developer perspectives. This translation is the starting point for the development of a consistent, common language.

Alignment and Sequencing

Figure 4 is intended to be descriptive, not prescriptive. The stages in the private sector and Federal processes may not always explicitly match and are shown graphically to reflect specific points of alignment as well as relative process alignment. For example, at the end of construction, both the Federal agency and developer will recognize a common point in time, represented here by the term "Acceptance" by the Federal agency and "Commercial Operation Date" by the developer.

The majority of stages are not shown to explicitly align because the stakeholders are driven by different goals; each stakeholder's stages do not coincide in time. Timelines for specific stages and descriptions of process are less rigid and not always aligned between parties, and so are shown this way in Figure 4. For instance, a Federal goal will be the release of a solicitation, whereas the developer's goal may be to obtain funding to respond to the solicitation. Therefore, the developer's milestone will not occur until after the solicitation has been released. Another area of difference is in the early stages of Market and Portfolio Analysis for the developer and the ID Opportunity Stage for the Federal agencies. These stages may have different lengths, with the Developer's stage shorter than the Federal agency's stage, as discussed below.

Limitations of Figures

Figure 4, and several variations of the same graphic, is used throughout this document as a tool to compare, at a high level, language and process across project participants. The figure is intended to represent, but not define, relative language and process stages of the parties. This is not intended, and cannot be used, as a representation or prescription of schedule or event timing for any given project type. To allow this high level contextual comparison, many simplifying assumptions were made – and some creative license taken, resulting in a simplified depiction of what can be a highly complex and variable multi-year process.

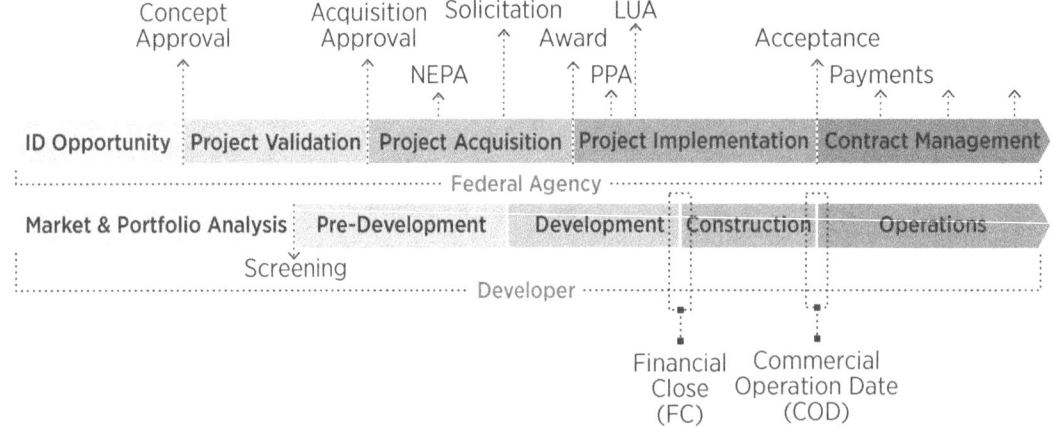

Figure 4. Developing a common language

II. A Reliable, Repeatable Project Development Process

This Guide establishes a common understanding around the fundamental principles of project development that can then be applied to both the Federal and private sector perspectives. The framework that emerges describes a project development risk management approach using an established, repeatable, disciplined process that is consistent with professional commercial practices and Federal requirements (including the Federal Acquisition Regulation). The general concepts of commercial project development and the project development framework are introduced in this section, with brief references to the comparable Federal process. In Section III (Application of Project Development by a Federal Agency), a Federal model that adapts the commercial process in more detail is highlighted.

The lifecycle of any project, Federal or private, begins with project development, moves to the construction phase, and transitions to the final operations phase. As a reliable, repeatable development process is developed, it is important to first establish its relationship to the other key elements in the project lifecycle, such as financial close and commercial operation date, as shown in Figure 5 (using terminology for phases that are generally understood by project developers).

The three elements of project development, construction, and operations are fairly universal, though the language within the Federal context can differ somewhat from the generic, commercial terms used here (see Section III, Application of Project Development by a Federal Agency, for more details). To allow the translation of terms from private sector to Federal sector and back, a more granular view of the project development stages will be discussed below and commercial and Federal terms will be translated so they can be used interchangeably. The activities included in the construction and operations phases, as defined above, will not be addressed as they are not the focus of this Guide.

Project Development Stages

To successfully attract private financing, a project must be fully defined with risks and unknowns mitigated and allocated to appropriate parties. For developers and providers of capital, judging the success of any project at its earliest stages depends on having a market, a way to get to the market, and the ability to obtain all of the relevant permits. The project development process ends with either an active decision to abandon the effort

Section II focuses on developing a process blueprint aimed at managing development risk and limiting financial losses from investments made in the development stage of a project or portfolio of projects, a key focus of the project developer. Understanding this, Federal agencies seeking to attract developers with private capital should focus on reducing real and perceived risks through the development and availability of data defining project feasibility and on providing transparency in both process and schedule.

or a successful project financing and the subsequent start of construction. The level of effort and investment required to fully define a project can be quite significant and, whether from the Federal or private sector perspective, this level of investment must be managed diligently through a rigorous process to protect resources.

For the purposes of this Guide, the project development phase is broken down into three stages: (1) Market and Portfolio Analysis, (2) Pre-Development, and (3) Development. These are generic commercial terms; those working from the Federal context may recognize terms such as (1) ID Opportunity, (2) Project Validation, and (3) Project Acquisition. These represent roughly the same activities with different naming conventions than the commercial terms mentioned above. A translation between these sets of terms is shown in Figure 6.

Depending on the method of financing used, early project development stages can be conducted and led by the sponsoring Federal agency, while the private sector takes the lead role during the project acquisition phase. Though this is a rough approximation of when the lead project development role switches from the Federal to the private sector, it is important for the agency to acknowledge that under certain methods of financing, the agency may incur the risk of losing its pre-development investment if the parties ultimately fail to reach an agreement. Consequently, agencies should seek to avoid investing in an acquisition process that generates a lackluster response by the private sector and can

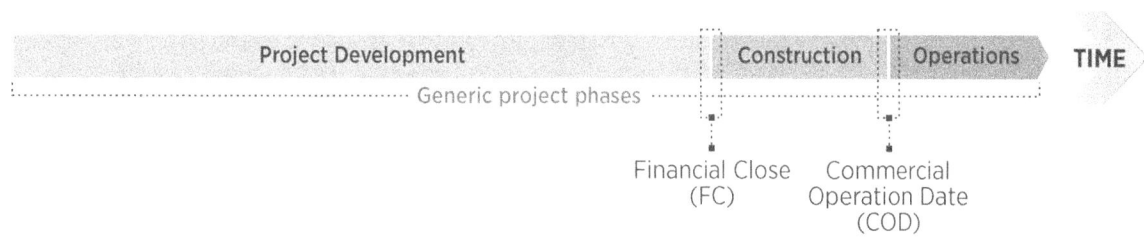

Figure 5. Project lifecycle phases

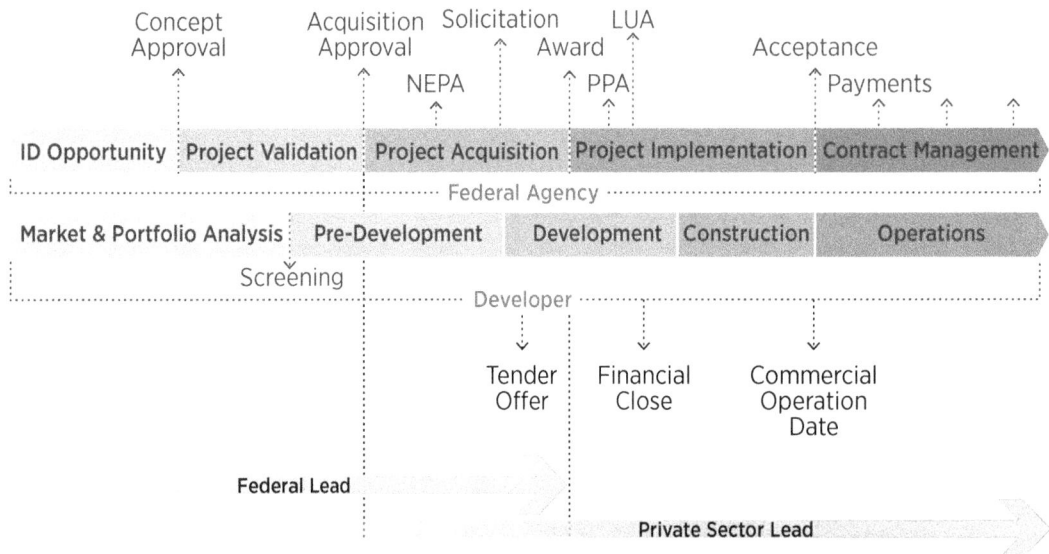

Figure 6. Project development stages, translated between government and private sectors [7]

do so by adopting the discipline, analysis, and decision making involved in the processes and frameworks provided in this Guide.

Stage 1. Market and Portfolio Analysis

The first stage of project development does not focus on a project, which may just be a concept at this stage. Stage 1 focuses instead on the market fundamentals that define or influence the project's operating environment. A project opportunity positioned in a market with supporting fundamentals has a strong economic business case, development and operational risks that are acceptable to all parties, acceptable technology or performance risk, site characteristics suitable for a given technology, supportive policies, and an execution pathway providing either access to markets or financing, or both.

> Translation: The private sector will commonly refer to this stage as "Market and/or Portfolio Analysis"; Federal agencies may use the term "Opportunity Identification," or something similar.

Developers identify market opportunities at several levels, and some of these levels are related to their core business model. By focusing on a set of technologies, or renewable resources, or other areas where the company may have a market advantage, the developers often have a shorter timeline in assessing a particular set of project opportunities than Federal agencies do. Some of the background work may have been done when setting up the company or developing the company business plan.

Federal agencies often view the planning and Market and Portfolio Analysis step as a new activity, which is sometimes not related to their core mission or area of expertise. The Federal agencies can take longer to identify opportunities, and this early stage can include significant planning at a regional or agency portfolio level. The agency also must review the potential projects despite limited familiarity with important facets of large renewable project development. While this process may seem long to the developers, it is critical to presenting sound projects for competition within the Federal system. This also introduces unique risk, as a project may face difficulty when market conditions change during the Federal planning process.

Strong project fundamentals and an understanding of how a project fits within a portfolio of opportunities are key foundations to the process. These provide the source of commitment and clarity of purpose necessary to both secure the resources required to develop a project and to persevere throughout the process. Without properly establishing project fundamental characteristics, the necessary resources (funding and skilled personnel) should not be allocated to move the project forward. A common mistake made by development teams in the early phases of project

> The discipline of avoiding, or abandoning low-probability projects as early in the development process as possible is a fundamental risk management function, and begins with market analysis.

---

[7] Figure 6 provides a general description of each party's responsibilities throughout the stages of project development. However, each party's responsibilities will depend on the type of financing employed (e.g., the private sector party has the lead role in the project development stage in the Energy Savings Performance Contract context). Please refer to the FEMP website for guidance on specific types of financing. http://www1.eere.energy.gov/femp/financing/mechanisms.html

development is to invest too heavily in technical or financial details before establishing that the project fundamentals are sufficient to maintain a sustained effort backed by common purpose and supported by leadership. Most developers will screen identified projects prior to moving into the next stage, moving only the strongest opportunities forward to the pre-development stage.

Elements of Project Fundamentals

FEMP has adopted the following five elements or categories from a NREL-developed framework to help organize the information required to establish sound project fundamentals.

Baseline

An objective analysis of the current energy market for the site that defines the market-based drivers supporting or motivating the development of the project. This analysis may consider the fuel source of the local utility, local or imported energy supply, existing or necessary infrastructure to support a project such as interconnection requirements, an assessment of competitive forces in the market, and the schedule of various factors such as incentives, transmission availability, and other market factors. Most Federal sites will consider Federal renewable energy goals and support for the mission(s) at the site, often including some element of energy security as motivation for a project. However, this Baseline analysis must go beyond Federal goals to consider the market context that will motivate a privately financed project. A clear summary of the analysis and project objectives will help the project move forward.

Economics

An objective analysis of fundamental energy economics must be established—both in terms of the market price of acquiring energy from existing sources (self-generated or utility-based) and from the proposed sources as comparison. If the proposed sources are likely to be more expensive, the differences must be acknowledged and dealt with upfront. Will the agency pay a premium for renewable energy? Is there some other value to be delivered by the project? Development and financing costs should be considered along with "overnight" construction costs [8] when considering project economics.

Policy

Policy and Execution Authorities must be addressed prior to expending significant resources pursuing a project. The Contracting Authority to purchase the energy and legal basis to provide the land to the developer must be clear. These can vary significantly between agencies and across the military services. Federal agency, state, local, and regulatory policy environments, including environmental regulations, must be examined for barriers to the project and steps should be taken to mitigate, remove, or deal with these policies to create the conditions for success. An assessment of local stakeholder support or opposition can be included in this category if information is available to gauge local support.

Technology

Fundamental technology assessment and analysis may be the most straightforward part of establishing project fundamentals. Engineering will be completed in Stage 2. In Stage 1, an assessment of available renewable resources and the commercially available conversion technologies to use the resource is essential to establish the likely reliability of the project's performance and gauge the investment community's willingness to finance it (bankability). This assessment should include a constructability review to establish fatal-flaw site constraints.

Consensus

Building from the Technology section, identifying key stakeholders (including local community and non-governmental organizations), and then communicating with and consensus-building among those project stakeholders are vital. To generate buy-in, a common understanding of the project's objectives and fundamental characteristics, and a unification of purpose are essential. Without consensus, staff and financial resources will not be made available, and stakeholders can become adversaries to the project when it is most vulnerable—before it gets off the ground.

Ideally, the end of Stage 1 in the Federal context will result in team consensus often demonstrated by a project concept approval or similar document.

Portfolio Analysis

In conjunction with an analysis of project fundamentals, or market analysis, a portfolio level view can be established. To achieve the highest return on the effort and resources expended to pursue large-scale renewable energy projects, a Federal agency should consider not only each potential project on its own merits of technical feasibility and market environment, but also the project within the context of the agency's portfolio of opportunities to choose the most valuable, feasible projects.

Federal agencies typically own and operate a portfolio of facilities and installations, with a wide range of size, geographic location, mission, and energy demand requirements. Each property has some technical potential for one or more renewable projects; for example, simply by virtue of being outdoors, the facility is subject to both solar and wind resources. Project economics are the next measure of feasibility; the constraint of energy cost is an important measure and may introduce a fatal flaw and direct efforts elsewhere.

For a more detailed discussion and an example of establishing a portfolio approach in the Federal context see Appendix A. Portfolio Approach.

---

[8] Overnight cost is an estimate of the cost at which a plant could be constructed assuming the entire process of planning through completion could be accomplished in a single day. This concept is useful to avoid any impact of financing issues and assumptions.

A private developer may not have easy access to information that would reduce cost and risk to early stage projects. Federal agencies should collect relevant information from previous studies such as NEPA documents and land use plans and make those available to developers. The Bureau of Land Management (BLM) has taken a proactive and coordinated approach by identifying land areas appropriate for Renewable Energy project development and developing "programmatic" assessments which pave the way for individual projects.

Stage 2. Pre-Development

The Pre-Development stage is meant to identify significant barriers to ultimate project execution prior to significant investment of time and money in the development stage. The goal of this stage is to uncover any fatal flaws with minimal investment of time and money and to confirm and establish project economics and the feasibility of obtaining all necessary agreements, approvals, permits, or contracts from third parties—without contracting or formally applying for them.

> Translation: The term "Pre-Development" is commonly used in the private sector for this stage; Federal agency employees may be more familiar with the term "Project Validation," or something similar.

Depending on the method of financing, early project development stages may be conducted and led by the sponsoring Federal agency. Agency leadership in the Pre-Development stage is important because the project at this stage is likely to be too risky to command an economic energy price, or perhaps any interest from the private sector at all. By performing some early development activities for the project, the agency can reduce the project risk. Lower risk will reduce the returns necessary for the developer and may lower the price of power for the Government.

Early development activities can consist of creating a financial model, or "pro forma," for the project to "run the numbers" and evaluate sector developers use their own proprietary pro forma analysis to assess these elements, and they apply their own risk tolerance and professional judgment to a project. Other activities may include:

- establishing that the site is available for development and transferrable to a private sector entity;

- producing a critical issues analysis ("CIA") report;

- confirming the renewable resource with site-specific data collection (solar and wind projects will generally require at least 12 months of data); and

- establishing a dialogue with potential off-takers or purchasers of the renewable energy produced by the project.

At the end of Stage 2, the project is likely ready to be offered to the public through a competitive procurement. Even projects with very strong project fundamentals require intensive data collection, analysis, and verification before a project becomes financeable and ultimately buildable. The data needed to create a government solicitation includes a large amount of this information.

Significant capital and investment of time by skilled professionals are required to develop a cohesive set of project data that will attract financing. Elements such as a development budget, permitting memo, tax opinion, and development plan for the full development cycle should be established in this stage to manage development cost and risk. These development costs may be an investment made at risk of significant or total loss—a risk that the private sector may be better equipped to take on. Accordingly, depending on the financing method, it may be more beneficial to both parties if the private developer assumes the lead role of the project at the end of this stage. To visualize this handoff of the project – and implicit transfer of the majority of development investment risk – to the developer, refer to Figure 2 in the Executive Summary.

In the Federal context, the successful completion of pre-development activities will likely result in approval of an acquisition plan.

Pre-Development activities are the beginning of the formal development process. The activities of this stage should be considered early steps within the same framework and approach as detailed in Stage 3, Development. Appendix B. Project Development Framework Categories lists pre-development steps for each of the framework categories described in Stage 3.

Thought experiment: Imagine being a developer and having a monthly meeting to discuss projects under consideration. Of the dozens of projects on the table, what ones get the majority of the attention and the commitments to action? Answer: Those with the most likelihood of being completed – those with the lowest risk.

Stage 3. Development

Once a potential project is found to have strong fundamentals in Stages 1 and 2, it moves into Stage 3, Development, in which the information needed to close a deal is generated, verified, and compiled as the basis of an executable transaction. It can be expected that developers of large-scale projects would require an off-take agreement or PPA prior to investing in Stage 3.

Managing the inherent risk of investing in this activity requires a regular, repeatable, documented project development discipline grounded in commercial project development practice paired with Federal procurement practices.

> Translation: The private sector refers to Stage 3 as "Development" to represent the largest commitment of time and money to prepare the project for financing and construction. Under certain financing methods, Federal agencies may use the term "Project Acquisition" or something similar to denote the transfer of the lead role to a private vendor.

In Stage 3, the investment required by the developer or Federal agency may increase dramatically as all the necessary documentation for the project is generated and negotiated by engineering, contract, and legal professionals preparing the project for financing and construction. This effort can entail significant resources (1% to 5% of total project costs), and can take from nine months to three years (or more). In Federal projects, this stage has two parts. One part is the detail developed by the Federal agency in order to issue a competitive process document and negotiate it through to acquisition award. The second part is the more detailed development work done by the project developer selected by the Federal agency to implement the project.

In the Federal context, this stage includes developing the RFP or other acquisition agreement (instrument), negotiations, and awards, and ensuring compliance with the Federal requirements for the type of financing method employed and other contract requirements. These requirements can include the PPA, land use agreement, and other activities that are critical to the developer's financial close. This stage is discussed in detail in Section III (Application of Project Development by a Federal Agency).

FEMP has adopted seven categories of information from an NREL-developed framework that can be used to organize and evaluate the risks and investment decisions required in Stages 2 and 3 of project development.[9] These categories form a framework of information on which an iterative process is conducted, supported by tools such as *pro formas* and development checklists or questionnaires. For more information on each of the seven categories below, see Appendix B. Project Development Framework Categories.

The seven categories are

Site

Site is the first element because a physical location for a renewable energy power project is required. An investor must be assured that he or she has access to the site for construction and operation of the facility for the term of the contract (Site Control). Federal agencies must also especially understand whether the site is affected by Bureau of Land Management (BLM) withdrawal terms, which affect terms of land use.

Resource

The renewable resource under consideration (sun, wind, biomass, or geothermal) needs to be characterized and understood at a level of detail and confidence appropriate to the project's stage of development. Whether the government or developer is investing in this resource data is an important consideration and can impact the viability and marketability of a project. Installation of measuring equipment and verifiable data collection can be costly and must meet lender and investor requirements; see Appendix B for more discussion.

Off-take

The off-take agreement is a PPA or other agreement that includes the terms of sale of energy between the project owner and the government, and any other characteristics of output of the project (such as Renewable Energy Certificates [RECs]) that generate funds to pay for the project.

> **Site, Resource, and Off-take categories are the core elements of project development because together they create value that promotes further investment. Securing these three elements by contract is a significant milestone for the project developer.**

This category also includes any necessary transmission access and related agreements necessary to get the power to the ultimate power purchaser; the terms of the PPA and other agreements must be established early in contract negotiation and ultimately, secured by a contract. For the Federal agency, the terms may be identified in the RFP or other document for the competitive process. In general, the rates proposed for the sale of power and RECs must meet the government's requirements, state laws, and procurement policy.

Permit

The permitting area encompass all permits necessary for project construction and operation—including all Federal requirements related to environmental regulations, such as NEPA, local electric utility interconnections, and necessary transmission rights or facilities. Permitting is an important element to understand from both a feasibility and risk standpoint—if a project has a high hurdle for permitting, and therefore includes significant risks, it needs to be considered with this in mind.

Technology

The technology area begins with the technical design feasibility of a given technology that was developed in earlier project fundamentals work and becomes more detailed through the project development process. This work culminates in the final selection of all technology vendors and manufacturers, securing quotes from Engineering, Procurement, and Construction ("EPC")

---

[9] SROPTTC™ is the project development framework, discussed herein, and developed at NREL. SROPTTC™ is a trademark owned by the Alliance for Sustainable Energy, LLC, the manager and operator of NREL.

contractors, selecting the team, and executing all supporting and related documentation such as warranties, guarantees, and performance requirements. Note that most large RE systems are designed to feed power only to an operating utility system. Special hardware and operational approaches are needed if the RE system is to contribute to energy security. These measures can increase costs.

Team

Early on, it is essential to assemble a qualified Federal team representing all aspects of the project including technical, financial, contracting, legal, real property, master planning, environmental, and operational aspects. Investors will look for a qualified and committed Federal team with requisite experience and capability.

Capital

Development capital is invested by developers to put all project development elements in place: Site, Resource, Off-Take, Permits, Technology, and Team. With every element documented, the project will attract the financial resources necessary for construction, commissioning, and initial operations. Raising and closing this financing is the final element in project development—but it is important to note that capital requirements do not begin at construction; they are required throughout the multi-year development process.

> **It is important to note that capital requirements do not begin at construction; they are required throughout the development process.**

Applying the Framework in a Project Environment (Considering the Financier's View Point)

Large-scale renewable energy projects need private capital to succeed. Financiers provide that private capital and have their own language and processes to navigate through project development. Ultimately, once a developer assumes the lead role on the project, a financier will likely be involved—the developer needs to assure he or she can attract an investor or financier, just as the Federal agency needs to assure it can attract a developer.

To assure Federal agencies develop projects that are viable to developers, Federal agencies need to understand how developers and financiers execute and invest in a project development process. This process, showing three distinct phases of activity

from the financier's perspective, with milestones that separate and define them, is shown in Figure 7.

The three phases can be summarized from the financier's perspective with an eye toward financial risk characteristics as follows:

1. The development equity phase represents the most speculative phase in which funds invested are at risk of total loss in the event a deal is not closed. Most of the development equity comes from the developer or other equity investors. Because this is the most speculative phase, debt is not available. Although equity investors will provide capital, they will expect a high return.

2. The construction finance phase represents the total capital cost of the project. Project financing incurs construction risks, but is mitigated by the creation of an asset, which is usually covered by some form of performance bonds or guarantees. Debt provided by the banking community is typically included at the construction phase. Because many of the risks have been mitigated at this point, debt providers, which are risk averse, can provide capital at a lower rate than equity investors. However, interest rates are still higher at this phase than at the next phase because of the reasons mentioned below.

3. The re-finance or permanent financing phase (also commonly called "take-out" financing) occurs when the speculative project has been converted into a stable asset generating income that is no longer subject to development or construction risks. Bank debt is also typically deployed as part of the permanent financing or re-financing phase.

The Financier's perspective and language can now be added to the translation diagram developed in Figure 4 to represent the terminology used by the financial community. Figure 8 shows this as a new horizontal band at the bottom labeled Financier. Third-party financing partners generally refer to the major phases of the project lifecycle in terms of financial risk. They may refer to the development equity phase, construction finance phase, and re-finance or permanent financing phase.

Risk in the Development Equity Phase

From the financier's perspective, different levels of risk tolerance and the skills and experience to mitigate those risks are

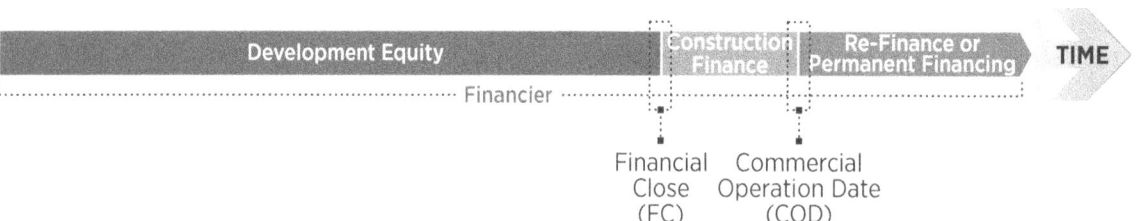

Figure 7. A financier's perspective of project lifetime and milestones

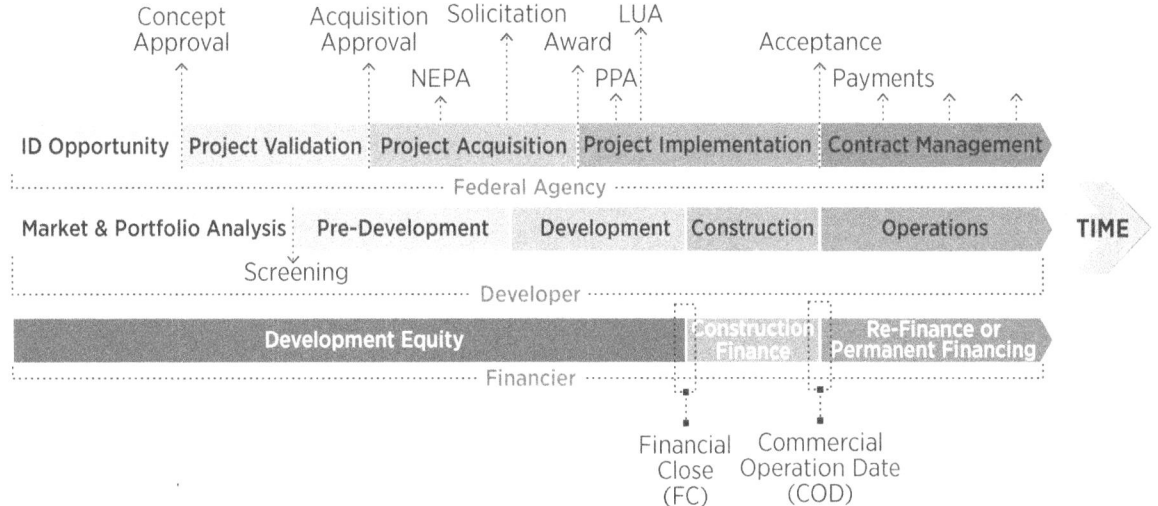

Figure 8. Process translation adding the financier's language

commonly found in the investors and lenders participating in funding each of these three phases. Different sources of capital may be employed to match the appetite for the risks and returns associated with a particular project lifecycle phase. Between the phases are two key milestones: first is financial close ("FC" in Figure 8), in which project financing transitions from development equity to construction finance. The second is commercial operations date ("COD" in Figure 8), where the project is considered fully constructed and ready for normal operations, and is therefore eligible for permanent financing.

Each phase of the project lifecycle has unique risk characteristics and the source of capital used reflects these different risk profiles. In general, as the project matures through the lifecycle, unknowns are steadily reduced as is risk.

The development equity phase does not follow this pattern; during project development, risk of loss moves counter to the decreasing unknowns. This occurs because any resources expended in the development equity phase can be lost completely if the project is not executed. As investments are made, the risk of loss increases along with the amount of money at risk – much of this investment may not be recoverable without a successful financial close.

Figure 9 shows a general risk profile across a project's lifecycle and financing phases. This profile does not parallel the declining trend of unknowns throughout the project lifecycle, but instead, the risk profile increases during the development equity phase. The project owner's perceived risk is actually increasing because each incremental dollar invested is subject to a total loss should a fatal flaw emerge prior to financial close.

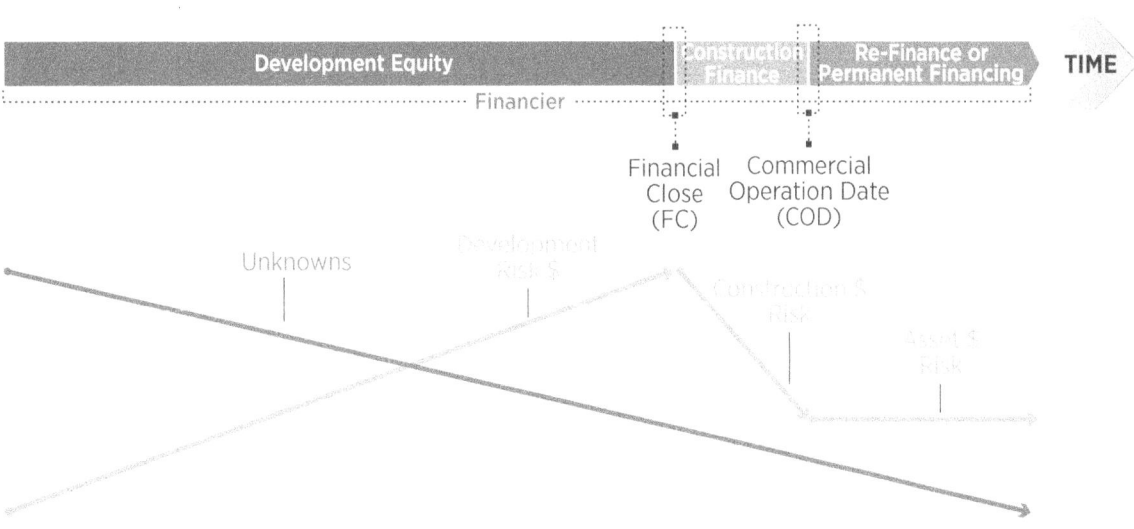

Figure 9. Project phases with risks and unknowns

It is this unique risk environment that drives not only the character of the capital sources used in the development equity phase, but also the process and management techniques used to mitigate risk and produce results. A high level description of a common approach is described next.

Iterative Approach/Incremental Investment

The nature of the risk profile for project development investments demands a process designed to mitigate the chance a project reaches 99% development before failing due to a fatal flaw, a situation with very negative financial consequences. Project developers, and Federal agencies taking the lead role in project development, can manage development risk using an iterative fatal-flaw analysis process.

A single iteration consists of confirming and documenting what is known in the areas of Site, Resource, Off-take, Permits, Technology, Team, and Capital, and then presenting that information in a format that can be used to inform a decision of whether to invest further in the project or stop in favor of other alternatives. Two tools are typically used to support this analysis: a development checklist and a pro forma. Samples of these can be found in Appendices B and F, respectively. A development checklist provides a basic list of issues to be checked and resolved, and typically evolves and grows with experience (new items are added from lessons learned on each project). A pro forma is a forward-looking financial model of the project and is used to forecast the results of the project development analysis in financial terms that can then be used to measure and evaluate the project's attractiveness throughout the project development process.

Figure 10 shows the major elements of a single iteration. Project development analysis is conducted by using a development checklist, translating project characteristics into a financial pro forma, and then assessing what was learned relative to the risks and rewards of the project. Each assessment asks the following questions:

- Has a potential fatal flaw been identified? If so, should the project be stopped?

- Have major risk areas been identified? If so, how can these risks be better quantified?

- Is the project still economically attractive? Review the budget on a quarterly basis.

- Where are the unknowns, and how can they be further mitigated?

- Where should the next dollar of investment in time and/or money be applied to reduce unknowns, mitigate risks, and develop key information?

After each iteration, the project is considered either viable and worth pursuing further or abandoned in preference for an alternative. If the project is considered viable, the most efficient way to increase the chances of success is through the elimination of unknowns by spending resources on the key areas of information (i.e., Site, Resource, Off-take (interconnection), Permitting,

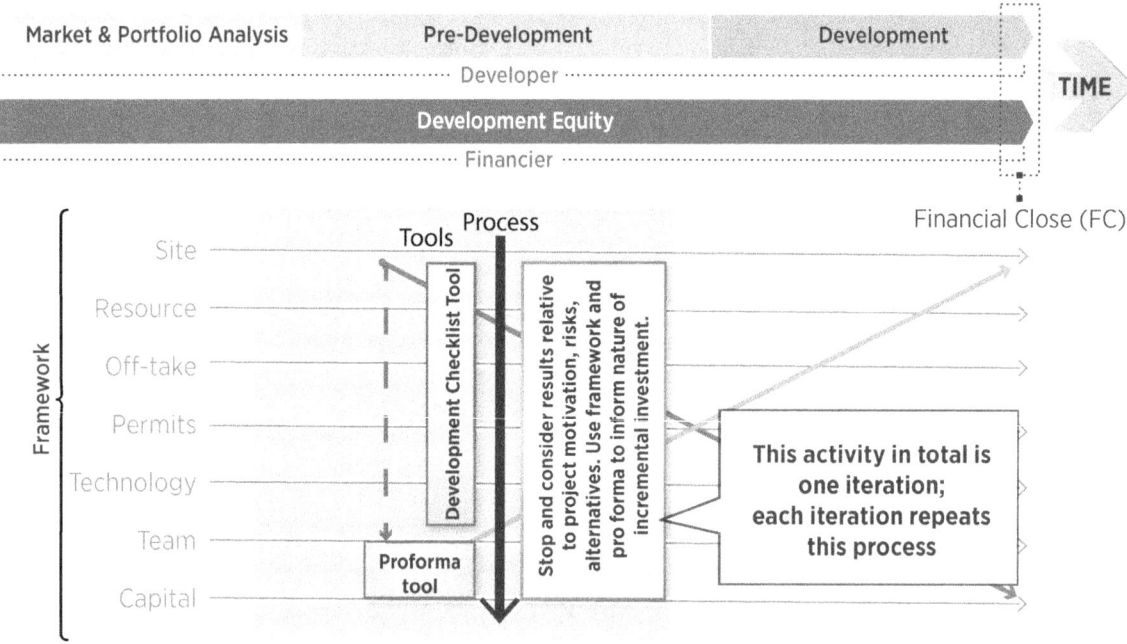

Figure 10. A single iteration in a project development framework

Technology (engineering), Team, or Capital). If a project is not considered to be viable and is abandoned, the process has worked and therefore has been a success. By identifying the fatal flaws of the project early, investment can be redirected toward the pursuit of the next best alternatives within the portfolio. This process is repeated iteratively, resulting in incremental investments and judgments amounting to a "Go Forward/Stop" decision each time. The result is a series of incremental investments, each followed by an assessment that systematically evaluates the project development framework categories, defines key parameters of the project, and drives unknowns from the system, as seen in Figure 11.

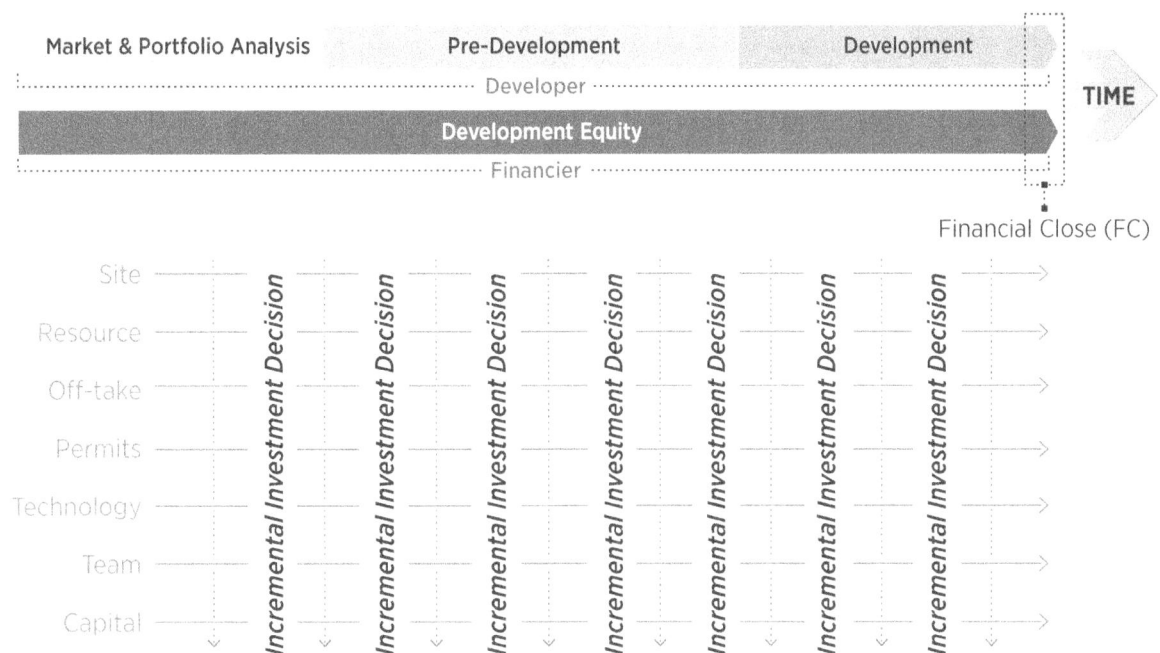

Figure 11. Iterations and incremental investment decisions lead to financial close

III. Application of Project Development by a Federal Agency

With the general concepts of project fundamentals and the project development framework introduced, it is time to consider these in action and apply them to projects developed in the Federal context. Through application of this guide, government staff will begin to understand commercial project development and financing, how the developer makes money, and the constraints of renewable power plant development. For more information on commercial project financing, see Appendix D. Commercial Project Financing.

In preparing this Guide, FEMP considered a wide range of government functions and experiences from energy and facility managers, Public Works officials, acquisition professionals, mission commanders, and Senior Executive Service decision makers. The Guide attempts to provide information to help each stakeholder navigate a complex transaction that is new to the Federal competitive procurement process. It does not attempt to provide comprehensive solutions or prescribe answers. It does provide a framework whereby a commercial process for renewable power plant development can succeed within the Federal environment. The framework and resources offered in this Guide, including checklists and a project risk assessment template and project validation report format from the U.S. Army, can be used as a guide to help ensure the success of Federal efforts and mitigate risk. However, these resources, found in Appendices B, G, and H, respectively, are not intended to be comprehensive, but to be examples, and should be used in conjunction with the support of qualified personnel.

> Projects that are presented to the private sector with weakly defined development risk and characteristics are unlikely to succeed. To accomplish the ambitious Federal goals with the highest efficiency, in certain methods of financing, Federal agencies may need to lead the risky early-stage project development activities (Market and Portfolio Analysis and Pre-Development stages). By taking on the role of the developer and financing the pre-development stages, the Federal agency effectively reduces the risk of the project and is able to present to developers a project that is better defined, less risky, and therefore, more likely to result in a successful completed project.

Applying the framework proposed in this Guide will help Federal project teams build strong business cases, define risks, and

establish good project characteristics that are attractive to the renewable investment community, helping the agency meet the significant goals and mandates to consume and produce renewable energy at Federal facilities.

To be attractive to the private sector, the Federal team should develop projects that have strong fundamental characteristics to the point where the private sector is willing and able to step in. Federal projects can be more attractive than commercial counterparts when the Federal Government provides land with good renewable resources, supports the permitting process, and purchases some, or all, of the output. This approach also benefits the Federal agency, as pricing for the project will be lower, because of the reduction of risk. It is important to execute a strong acquisition strategy that demonstrates these project characteristics to potential developers.

It is also important that the government and developers are able to demonstrate success early in this new industry. This requires a common understanding between the parties. Many of the faults found in past Federal contracts related to renewable energy can be attributed to a failure by the government to adequately understand the commercial power plant development side of the transaction during negotiation. The principles in this Guide will enable both sides of a transaction to better understand the deal because better informed people execute better deals.

Projects that are presented to the private sector with weakly defined development risk and characteristics are unlikely to succeed. To accomplish the ambitious Federal goals with the highest efficiency, in certain methods of financing, Federal agencies may need to lead the risky early-stage project development activities (Market and Portfolio Analysis and Pre-Development stages). By taking on the role of the developer and financing the pre-development stages, the Federal agency effectively reduces the risk of the project and is able to present to developers a project that is better defined, less risky, and therefore more likely to result in a successfully completed project.

The techniques discussed in this Guide also benefit the government approvals and acquisition process. Applying the due diligence techniques will create a thorough set of information to facilitate government decision making. The incremental development steps help to ensure stakeholder issues are identified and addressed throughout development of the project. Government development staff will be able to demonstrate good projects for leadership approvals, secure funding for project development stages, develop strong solicitations, and negotiate commercially viable deals benefitting all parties.

Making Federal Projects Attractive to the Private Sector

With the help of this Guide, private sector developers can begin to understand how to operate within the constraints of the Federal agency energy development and competitive acquisition environment. The paradigms for doing business in the renewable energy industry are continuously evolving, making it more difficult to fit within the Federal process that prefers business methods that are mature and stable and that must work within applicable laws and regulations.

The Federal sector must be an attractive transaction partner to persuade the renewable developer community to mobilize the skills and private investment necessary to move projects forward and to achieve agency goals. The entire renewable energy industry is competing for a very limited pool of development funds; thus, opportunities that are presented to the private sector with weakly defined development risks or characteristics are unlikely to succeed.

The Federal Government can also leverage its credit as an important, and attractive, attribute. To a developer, a long-term agreement with the U.S. Federal Government should be a low-risk revenue stream that will garner a corresponding lower cost of financing. This provides a great opportunity for the Federal Government to reduce the cost of energy for the government and provides stable, long-term energy costs.

In order to share the risk with the developer and further a project, the government may pay for studies that the government requires to comply with government procedures. Examination of previous projects that were unsuccessful could result in "lessons learned" and can inform decisions about how a project would have to proceed in order to be successful.

As noted previously, the principles herein must be applied with caution and with the support of qualified experts. A comparable example is an aircraft flight manual. A few people may be able to read the manual and successfully take off for a flight, but it is likely to be too risky. The recommended approach is to use the services of a qualified flight instructor. Likewise, successful application of this Guide will require expert support by subject matter experts (SMEs) experienced in renewable energy project development. This subject matter expertise and the function of directing a project or portfolio of projects can come from a team of Federal employees or from contractor support. As discussed in stage 3 of this section, industry should be aware that contractors involved in setting requirements for a specific project are generally not

eligible to bid on the resulting acquisition. Regardless of the source of expertise, it is paramount that the government funds the SME function to develop projects that will be privately financed. As it is for any significant transaction, it is important for large energy projects to establish complementary resources on both the government side and the developer side of the deal. Employing SME resources will support projects and position the Federal Government as an attractive sector within the large-scale renewable energy development marketplace. Limited resources to help with the project development process can be accessed through the FEMP website at www.femp.energy.gov.

The Federal Process

It is important to be able to match recognizable stages of Federal project management with the stages of commercial project development described in Section II (A Reliable, Repeatable Project Development Process). The Federal process occurs in five sequential stages that are similar to the five stages of the commercial process as shown in Figure 12 below. Each agency will customize individual process elements to conform to its own policies, regulations, and applicable laws. (Actual examples drawn from the Army's Energy Initiatives Task Force are provided in Appendix G. Project Risk Assessment Template and Appendix H. Project Validation Report (DRAFT). A clearly defined process for determining potential impacts that may constitute a fatal project flaw is extremely important to ensure reasonable expectations for all parties.

The early government stages parallel the commercial stages as shown in Figure 2 earlier in the Executive Summary of this report. Stage 1 "ID Opportunity" and Stage 2 "Project Validation" are comparable to the commercial Market and Portfolio Analysis and Pre-Development stages. By taking on the role of the developer and financier for certain types of financing during these early stages, the Federal agency can use the techniques described here to methodically choose projects that are more likely to be successful when offered to commercial developers. These techniques effectively reduce project risk, allowing the government to either present a project that is better defined and more likely to result

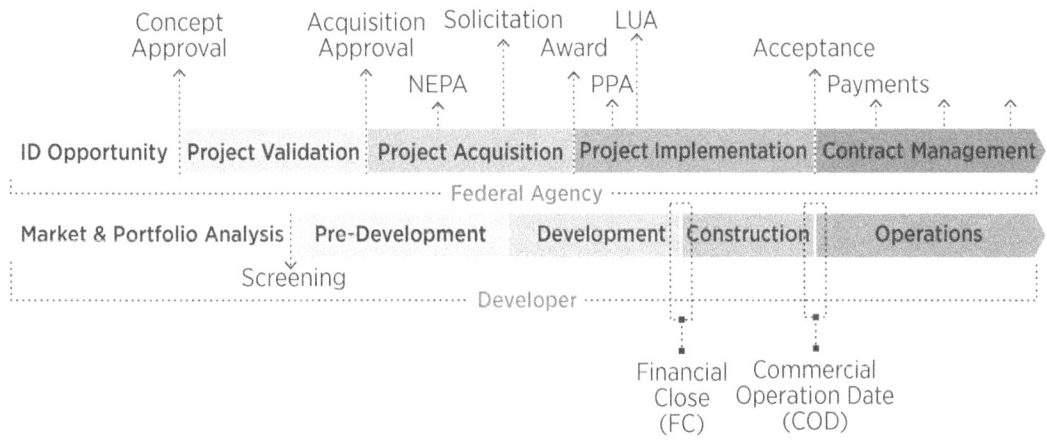

Figure 12. Project stages

in a successful project or to disengage because the project is flawed. Ending development of poor projects can be challenging, but it is vital. Government processes should encourage identification of flaws and applaud decisions to stop, so that good projects can thrive and have the resources needed to complete a successful competitive process.

Figure 13 below illustrates certain financing implications in which the government assumes the lead role in pre-development stages within the project development framework, taking on the early risk before transferring the project to the developer at the contract award.

The Federal pre-development investment will be personnel time, direct costs for studies, environmental management, and legal services. The techniques in this Guide help Federal personnel to continually assess the likelihood of success, decide how much funding to put at risk, and when it is appropriate to withdraw from a risky project.

Stage 1. Identify Opportunities

A portfolio approach is recommended to optimize the allocation of resources to meet individual agency goals, an approach resembling many planning processes in Federal agencies. Its purpose is to validate the requirement, assess the ability of the agency or service to meet the requirement, identify resource needs (funding, staff, and land), and select the best opportunities to work on. More information on the portfolio approach to project development is presented in Appendix A. Portfolio Approach.

The process for identifying the best project opportunities that can succeed in the commercial market is described in Stage 1 Market and Portfolio Analysis in Section II (A Reliable, Repeatable Project Development Process). This section also discusses project fundamentals. In Stage 1, a high-level analysis of project fundamentals and an understanding of how a project fits within a portfolio of opportunities are key foundations for choosing projects to pursue. These fundamentals provide the source of commitment and clarity of purpose necessary to both secure the resources required to develop a project and to persevere throughout the process. Without properly establishing project fundamental characteristics, including team consensus on the project, the necessary resources (funding and skilled personnel) should not be allocated to move the project forward.

Federal project analysis should follow a process similar to that used by private developers, with specific agency criteria added. As discussed in Section II, Federal agencies can take a long time to plan, identify, and vet a potential large renewable project or set of projects. This process can include coordination with real estate and mission commanders for conflicting land uses, master planning for long-term availability for the land, and many other factors. The steps identified with project fundamentals, as well as some pre-development and development tasks, can take considerable time. As noted in Section II, this process is often longer than that used by private developers. However, completion of these tasks improves the likelihood of successful, competitive solicitation and development stages.

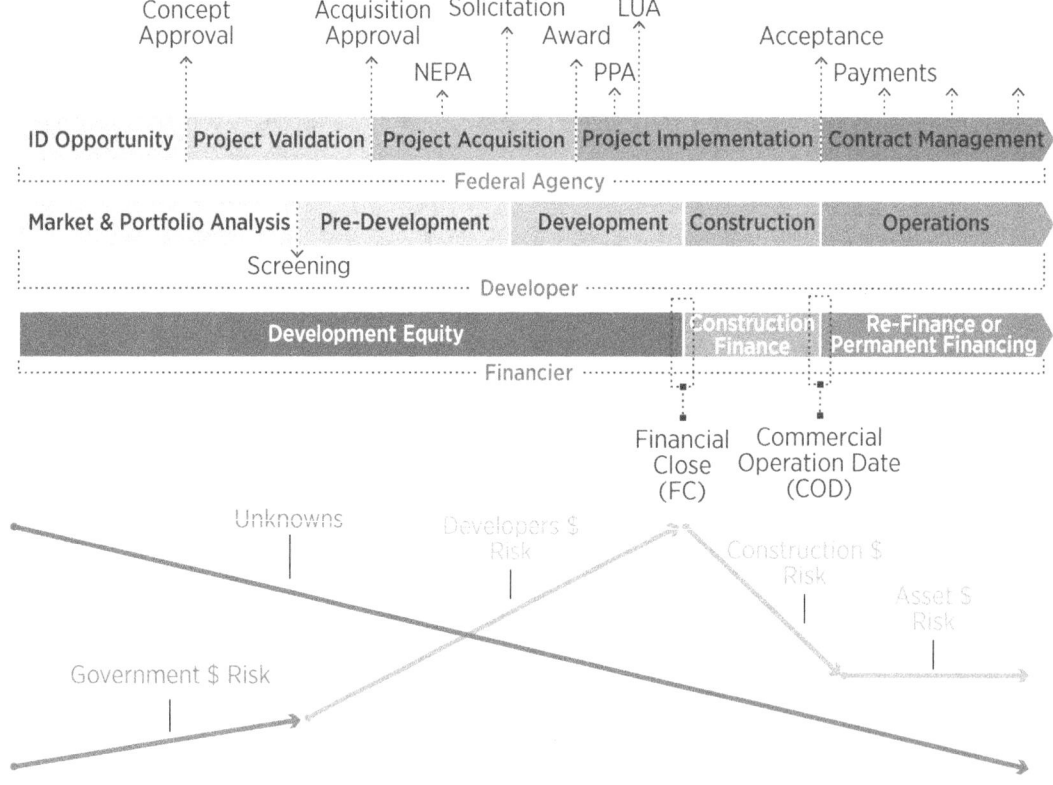

Figure 13. A typical government process for certain types of financing

When Stage 1 is complete, agency leadership will be able to approve projects and development tactics based on strong fundamentals defining opportunities for a successful project that will meet Federal and agency goals for renewable energy, GHGs, cost, and energy security. The result of this stage is the conceptual approval from senior leadership for each of the projects to be developed.

Stage 2. Project Validation

This stage parallels the commercial pre-development Stage 2, detailed in Section II (A Reliable, Repeatable Project Development Process). The pre-development stage is meant to identify significant barriers to ultimate project execution prior to significant investment of time and money in the development stage. Developers use their experience and base knowledge to add their judgment to this process; Federal employees must do the same, possibly with the support of subject matter experts. The goal of this stage is to uncover any fatal flaws with minimal investment of time and money, and to confirm and establish project economics and the feasibility of obtaining all necessary agreements, approvals, permits, or contracts from third parties—without contracting or formally applying for them. Government stakeholders in a project cooperate to achieve consensus on the goals, steps, and criteria for success. The key project characteristics are established in accordance with industry (and financier) standards so that the project development risk is clear. More detailed analysis of fundamentals that determine the likelihood for success is established in this Project Validation stage. See Stages 1 and 2 of Section II (A Reliable, Repeatable Project Development Process) for details. Although the timeline for the developer stages and the government stages do not line up exactly in Figures 12 and 13, the steps and considerations are very similar. In addition, detailed items to be considered during Stage 2 are listed under pre-development in each topic category of Appendix B. Project Development Framework Categories.

It is important to ensure that the development team represents key interests of a comprehensive stakeholder group including the energy manager, department of public works director, base commander/site director, senior mission commander, contracting officer, acquisition team, legal counsel, base operations, real estate and master planning personnel, agency/service leadership, and environmental experts. It is strongly recommended that the local utility is also closely involved to ensure that the proposed project is viable within the constraints of the applicable laws and regulations.

A sample Project Risk Assessment Framework used by the Army's Energy Initiatives Task Force can be found in Appendix G. Project Risk Assessment Template.

Even projects with very strong project fundamentals require intensive data collection, analysis, and verification before a project becomes financeable and ultimately buildable. The data needed to create a competitive solicitation includes a large amount of this information. In addition, this data should match the data quality standards of developers and financiers, lest developers back away from a solicitation.

As the government agency proceeds through this stage, decisions must be made concerning how far certain elements will be taken by the agency and which elements the selected developer will be responsible for. As an example, resource data collection is a vital element of the project involving significant investment, expertise, and requirements to be met. Consideration must be given as to which party (government agency or developer) is best equipped to manage this investment.

NEPA Considerations

An important component of the Project Validation stage is the NEPA process. Compliance with NEPA is a Federal responsibility. The Federal agency is responsible for the quality of data and analysis in the NEPA review and any subsequent decisions. The role of a developer in the NEPA process may vary within appropriate limits, including schedule. For the projects in this Guide, the NEPA process will usually occur during the project acquisition stage, prior to the release of the solicitation, because the agency has a very well defined project—including project size and location—for which the agency is issuing the solicitation. Sometimes the project will be defined by the project developer, and in this case, the NEPA process would start after the solicitation/selection of the project and be completed prior to a final agency decision whether to approve the project.

The Federal agency always manages the NEPA process and issues decisions. The developer may pay costs for preparing the NEPA review, will provide at least some of the data needed for the analysis (e.g., information about the proposed project), and may have other roles depending on the circumstances. The project developer does not, however, control the process. The NEPA process aims to insure that the agency takes a "hard look" at the environmental consequences of a proposed action and to make information on the environmental consequences of the proposal available to the public. The heart of the NEPA process is the exploration and evaluation of a range of reasonable alternatives for agency decision making.

Many factors evaluated during the NEPA process, including alternatives to the proposed action and potential mitigation strategies, can affect the project's ability to be competitively developed and financed in the private markets. The intent here is to highlight the importance of defining the scope of the project and information regarding the environmental consequences of the project and reasonable alternatives in order to understand how addressing the environmental consequences may impact the project's ability to be competitively developed and financed in the private markets. As a result, for these privately financed projects, agencies should consider ways to gather information about the environmental impacts and the financeability of the project and its alternatives either before or early in the NEPA process. This information can be used in further developing the scope of the project and its alternatives that may have a higher potential for successful financing of the project. In addition, the agency should request and consider developer input on the financeability of alternatives compared in an Environmental Assessment (EA) or Environmental Impact Statement (EIS) to fully inform the final agency decision. The reader should also refer to and consider comments relative to NEPA in the following section titled "Stage 3; Project Acquisition" and in Appendix B4, "Permits including NEPA Compliance and Permitting Activities."

Upon completion of this Project Validation stage, the government is likely to have assembled a complete set of data to enable approval of an acquisition strategy for a competitive acquisition process such as an RFP, and will have secured senior leadership approval to proceed. Key elements of securing this approval will include forecasts of the project energy costs, an acquisition strategy, impact on the mission at the site, and early stages of NEPA compliance considerations. If the project is found to have fatal flaws, the project may be stopped and the agency will be free to move on to the next potential development project. If the project is approved, the Federal agency typically will transfer the project to the private sector through the acquisition process, because further investment by the agency is likely to generate diminishing returns.

Stage 3. Project Acquisition

Once a potential project has an approved acquisition plan, it moves into Stage 3, Project Acquisition. This stage starts the engagement with commercial developers on individual projects through a competitive process, usually an RFP, executed under the supervision of a contracting officer in accordance with the acquisition strategy approved in Stage 2 above. Federal contract rules impose requirements that may be different from industry standard practice with private projects. An important Federal contracting principle is that firms that help develop requirements for a specific project are generally not eligible to bid on the acquisition resulting from those requirements, absent a written waiver of the conflict of interest. However, firms usually may provide general information on technology, market conditions, and other relevant information without a conflict arising. In addition, a developer that is eligible to bid on an acquisition can include in the price of the proposal the value of any initial work the developer puts into a project prior to winning the award. It is possible that the contracting vehicle selected may limit the pool of eligible developers. Federal agencies and private sector developers must both remain aware of these limitations.

In Stage 3, the information needed to prepare the documents for a competitive acquisition is generated, verified, and compiled as the basis of an executable transaction for both the government and the developer seeking financing. Managing the inherent risk of investing in this activity requires a regular, repeatable, documented project development discipline grounded in commercial project development practice paired with Federal acquisition practices.

In this stage, the investment required by the Federal agency may increase significantly as all the necessary documentation for the competitive process for the project is generated by engineering, legal, acquisition, and other professionals. The steps include preparing to issue a competitive solicitation, evaluating responses to the competitive solicitation, negotiating a contract award, and documenting the contract. Stage 3 also includes the financing stage of the project, which generally begins after the development is complete and can demonstrate that it is ready for execution, and prior to start of construction. The developer is responsible for this effort, though the Federal agency will likely be involved as a key project stakeholder and a party to key agreements, so agency personnel and SME involvement will still be necessary.

The government must also address the requirements for connecting the proposed system to the electrical grid. Even when within the "fence line," important safety and reliability requirements and interconnection standards must be met to connect generation sources to the grid. The utility will do most of this work, providing a well-documented set of requirements for interconnection, which can be included in the solicitation to provide detailed information and minimize cost estimates. For a large-scale project providing power to the site, the Federal agency may also be required as a party to an interconnection agreement as the account holder. The work done previously in Stage 2 and in this acquisition stage will equip a government source selection team to review the solicitation responses against a set of selection criteria, negotiate, and conclude the acquisition stage with a contract award.

For the Federal agency, the Project Validation stage and this acquisition stage effort can entail significant resources (from 1% to 5% of total project costs).

As in the Project Validation stage, the early part of the acquisition stage will parallel the development process described in detail in Stage 3 of Section II, and detailed in Appendix B. All seven categories should be analyzed, and the first three categories of Site, Resource, and Off-take, plus NEPA compliance and permitting activities, should be addressed in acquisition documents.

NEPA Considerations

The government should address several major power plant development issues during this development stage to reduce the risk of cost and schedule overruns. The first is NEPA, which can be an expensive and time-consuming process. Compliance with NEPA is a Federal obligation that cannot be delegated to private parties and should be integrated into the project planning process to ensure that planning and decisions reflect environmental considerations so that delays can be avoided later in the process. Agencies should develop meaningful and expeditious timelines for environmental reviews and should work in close consultation with developers to gather data efficiently and cost effectively. When possible, NEPA reviews should be coordinated with other permitting and review processes (e.g., consultation under Section 106 of the National Historic Preservation Act), so that reviews can be accomplished concurrently and collaboratively, rather than sequentially. Agencies should recognize that any mitigation that the agency may require to avoid or reduce adverse environmental impacts could affect the technical or economic viability of the project itself (e.g., by altering the design or cost of the proposal). The Federal agency and developer should discuss mitigation measures and options as early in the NEPA process as possible so that the developer can make comments and can adjust its plans accordingly. As stated earlier, the intent is to highlight the importance of these decisions to both the agency and the developer—all decisions impact a project's ability to be competitively developed and financed in the private markets. The reader should also refer to and consider comments relative to NEPA in the previous section titled "Stage 2; Project Validation."

> The Federal agency is buying electricity from the developer, not procuring construction services and equipment; the developer/Federal relationship is fundamentally different when the project is financed by the developer.

The permitting that occurs in Stages 2, 3, and 4 of this process framework is central to the Federal role in a project to deliver renewable energy to a utility or agency. A clearly defined and repeatable permitting process reduces the risk and cost of the developer and results in more successful projects. The value of Programmatic EIS efforts for wind and solar on BLM land is recognized.

Stage 4. Project Implementation

Once a developer is selected by competitive process award, the Federal role changes to support and monitor the developer's implementation plan. This is a significant change from the traditional Federal role of reviewing and approving designs and managing construction. Generally the Federal Government will be primarily concerned that the developer can deliver and operate the project within the standards defined in the contract. Key government elements in the process include execution of the land use agreement and completion of the terms of the PPA (contract execution). The rest is up to the developer.

During this stage, the developer completes the project financing, design, and permits for construction, thus completing the developer's Stage 3. Once the developer completes the financial close milestone, construction begins. For Federal agencies this implementation stage includes construction, which is a separate stage for the Developers and Financiers. All the stages converge again at the Commercial Operation Date ("COD"), which may coincide with any necessary Federal acceptance of the project.

Stage 5. Contract Management

The process moves from implementation into management once the power plant has been commissioned and is online. The commercial term for this is Commercial Operation Date. At this point, the developer has met the requirement to build a power plant capable of operating at contract outputs. Thereafter the developer operates and maintains the plant to continue to produce energy at the contracted levels. The government ensures that the quantity and quality of energy meet the specifications and pays for that energy. To minimize the risk of a project failing at a later stage, the government is likely to require regular review of the operations, maintenance, and capital reinvestment plans of a project, but only to the extent that operational problems are affecting the generation and delivery of the contracted energy, and in line with project contracts and agreements.

Conclusion

For certain financing methods, the Federal agency's execution of the first stages of this process and the developer's assumption of the later stages must be coordinated throughout Stage 3, Project Acquisition, to ensure success. The probability of failure escalates if the guidelines proposed in this document are not applied. In that same vein, if the Federal agency inadequately prepares a solicitation used for the acquisition, the results may include:

1. Poor bid responses because of lack of interest from developers and financiers. This results in a waste of time and money since no contract awarded.

2. Projects will be successfully awarded, but implemented poorly or not at all. This outcome results in wasted Federal and private sector investment. Goals will not be met. Investment will have been wasted on a project that cannot be executed because the procurement allowed the selection of an inappropriate developer that fails to perform. The government is exposed to additional costs to replace the energy that is not delivered.

A comprehensive approach to defining and pursuing good project opportunities is essential to establish and maintain a track record of success that attracts investment. When development risk is not managed or managed inefficiently, it manifests itself in the failure of later-stage projects and financial losses. Losses can quickly add up; investors (tax payers or the private sector) tend to have long memories of losses, impairing the ability of projects in the Federal sector to be executed at all.

The question of roles and timing of the transition from Federal to private sector investment does not have a simple, universal answer. There will be unique aspects to most projects that will drive specific decisions. A general approach is offered in Figure 14 below. The figure shows a transition point within the developer's Stages 2 and part of Stage 3, with the Federal sponsor developing project fundamentals to mobilize the resources required to get through the pre-development stage. The transition occurs as the developer validates the Federal project data and submits a proposal in response to a solicitation. After evaluation, negotiation, and contract award, the developer has the primary responsibility for completing the project. The Federal Government continues its ongoing participation as a transaction counter-party (meaning the longer-term commitment as a party to a long-term agreement such as a PPA) after the competitive process award date. This acknowledges that a contract award does not define the end of the Federal role. Ongoing resources and expertise will be necessary to provide a strong and effective partner throughout the remaining development process. Appendix E. 10-Step Project Development Framework Approach provides an example approach to developing large-scale renewable energy projects with the developer's project development framework aligned with the government stages.

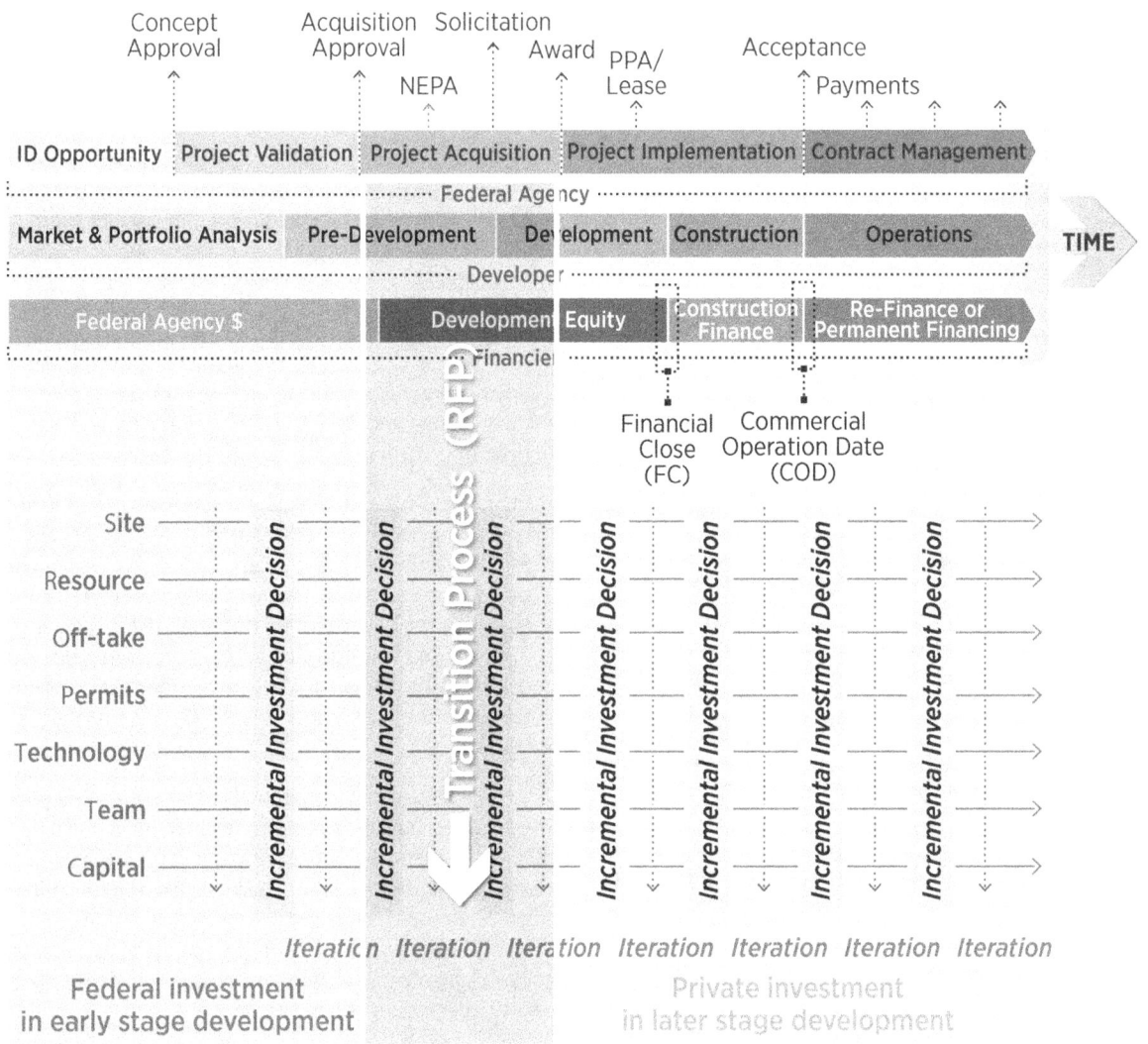

Figure 14. Transition between Federal sector role and private sector role across project lifecycle

IV. Outlook

The success of the Federal renewable energy market depends on the ability of agencies and the private sector to recognize each other as essential to reaching a common goal. Neither party will be successful if the requirements of each are not met and constraints are not overcome. The methodologies of each party must be translated to each other so that a common language and purpose is developed and maintained between all parties.

Both parties should keep in mind the following areas as the Federal market for renewable energy development continues to develop and mature:

- Opportunities for financially viable projects exist today in the U.S. that meet the requirements of both Federal agencies and the private sector.

- A predictable process is likely to improve the Federal sector's ability to attract private capital to Federal projects. Contracting forms, process steps, and schedules that are certain and predictable will generate significant investor interest and drive developers to compete for Federal projects.

- Once acknowledged, project development risks can be managed and should not deter the pursuit of projects that meet the requirements and needs of the Federal Government.

- Financing is available for renewable energy projects, subject to the competitive nature of capital markets that seek the highest risk-adjusted returns. Renewable project opportunities absolutely must remain competitive within the broader market or the projects will not move forward.

- Although this Guide focuses on large-scale renewable energy opportunities, smaller-scale distributed energy applications are often an opportunity Federal agencies can benefit from. The same principles provided in this Guide are applicable to smaller projects.

V. Points of Contact

Anne Crawley
U.S Department of Energy
Federal Energy Management Program
202-586-1505

Boyan Kovacic
U.S. Department of Energy
Federal Energy Management Program
202-586-4272

Andy Walker
Senior Project Leader
Integrated Applications Center
National Renewable Energy Laboratory
303-384-7531

Additional Resources

FEMP Renewable Energy Project Assistance
http://www1.eere.energy.gov/femp/technologies/renewable_assistance.html

FEMP Funding Options Website
www.eere.energy.gov/femp/financing/mechanisms.html

FEMP Glossary
www.eere.energy.gov/femp/information/glossary.html

FEMP Project Funding Quick Guide
www.eere.energy.gov/femp/pdfs/project_funding_guide.pdf

FEMP Authorizing Laws and Regulations
www.eere.energy.gov/femp/regulations/regulations.html

Appendix A. Portfolio Approach

To achieve the highest return on the effort and resources expended to pursue large-scale renewable energy projects, a Federal agency must consider not only each potential project on its own merits of technical feasibility and market environment, but also these things within the context of the agency's portfolio of opportunities to choose the most valuable, feasible projects.

Federal agencies typically own and operate a portfolio of facilities and installations, with a wide range of size, geographic location, mission, and energy demand requirements. Each property has some technical potential for one or more renewable projects; simply by virtue of being outdoors, the facility is subject to solar and wind. Project economics are the next measure of feasibility; the constraint of delivered energy cost is an important measure and may introduce a fatal flaw and direct effort elsewhere.

In this discussion of using a portfolio approach to project development, there are two concepts assumed to be understood and must therefore be introduced before going forward:

Constraint: Limited Resources

Every agency has limited resources; developing projects takes experienced, knowledgeable human resources as well as operating budget for direct costs. To apply these resources efficiently, the question is asked: With technical and economic feasibility established at a particular site, how does the level of feasibility compare with all other feasible opportunities?

Goal: Maximizing Returns and Meeting Federal and Agency Goals

As mentioned earlier, some measure of energy output, financial return, or other benefits in exchange for the resources invested in pursuit of renewable energy projects is expected. Whether it is an increase in renewable energy use or production, a reduction of GHG emissions, the diversification of energy supplies, the completion of mission requirements, cost savings, or energy security, the investment in renewable projects has a purpose and the impact or contribution must be measurable.

The goal of maximizing returns is most easily associated with private business. Government entities measure the effectiveness of expenditures not in the form of profits, but by yield on the investment toward a specific purpose or requirement. In the case of renewable energy project development, regardless of purpose, the yield is achieved through completed projects. The investment is the time and money invested by the government to make the project happen. The achieved purpose may be something that the renewable energy satisfies, for example a renewable energy goal or GHG reduction. If deploying renewables reduces an agency's GHG footprint, and therefore projects are pursued, the scale to which the purpose is served will depend on metrics such as the megawatt-hours produced by the renewable power plants, and therefore the offset of emissions from fossil sources.

Portfolio Then Project

Pursuing a renewable energy project without first developing an options portfolio can put an agency at a disadvantage. Moving a large-scale project from concept to reality will require the focus, sense of purpose, and commitment generated from the knowledge that there is simply no better opportunity to pursue. The project itself will benefit greatly from this motivation; as challenges arise, the motivation will help to overcome them.

Project development deals with uncertainty; very little is known when the concept for a project is first developed—by the time it is completed and operating, most everything is known. Developing a portfolio-level understanding of renewable opportunities serves this function well; knowing that some measure has been taken to prioritize the project removes the uncertainty that it may be the wrong one.

Portfolio Analysis Steps

Federal agencies can approach a portfolio of facilities, installations, or land holdings with a systematic approach to begin to identify leading candidates for large-scale renewable projects, and then focus resources and gain experience with technically and economically robust projects.

1. Establish energy demand – Monthly and annual usage, unit costs, and future growth trend for each facility.

2. Technical feasibility – Using publically available data from NREL and others, rank sites based on the resources available. Rank solar sites, from best to worst, and so on for each resource type.

3. Economic feasibility – Renewable energy will compete with either the retail cost of electricity (behind the meter installations) or the wholesale market (utility scale); review state policies for market structures like regulated vs. unregulated markets, and Renewable Portfolio Standards (RPS) including incentives, which will have strong impacts on economic and market feasibility. Check local policy environments for local incentives that may provide economic advantage.

4. Market feasibility – Because of the nature of electricity, market and/or physical constraints can limit a technical/economic project from "getting to market." Examples are a lack of supportive policies (e.g., feed-in-tariffs and net-metering), market structures (e.g., regulated vs. unregulated markets), and a physical pathway through transmission infrastructure. Because of influences like this, going beyond technical and economic feasibility analysis is required to establish the ultimate viability of a project.

5. Ranking and analysis – With steps 2 through 4 above completed for each technology (i.e., solar, wind, and other renewables as defined on page 2), the basic data is in place to begin ranking technical and economic (and completion) feasibility across a portfolio of sites.

Strategy

With portfolio analysis and ranking in place, the total list of potential projects can be integrated, and the top-ranked sites pursued. Deployment goals with near and long-term metrics are helpful to help guide decision making and strategy. Examples might be a strategy that recognizes and balances output metrics like megawatts installed or megawatt-hours produced in terms of both annual and long-term cumulative goals.

The element of time will play into strategy, as some technologies take longer to develop than others. Balancing near-term renewable energy production with longer-term goals may suggest pursuit of, for example, solar and geothermal projects; this allows solar projects to come in faster (but with lower output per installed megawatt) and geothermal to contribute to a production portfolio later on.

Competition and Discipline

The element of competition comes into play when operating in a portfolio context that is subject to the concepts introduced earlier: limited resources and the goal of maximizing returns. Resources are limited, so not all projects can be served; competition for those resources is a healthy mechanism to operate as efficiently as possible with respect to yield per unit of investment. In other words, competition is necessary to maximize returns.

Competition

Competition manifests itself in the investment decision. As an example, a hypothetical generic Federal Agency (FA) with 100 geographically diverse sites or installations has performed a portfolio analysis as outlined above and now FA has a list of priority projects, ranked 1 to 100 in order of priority. Engaging in any one of these projects will begin to consume resources; however, it is not yet known if the project will ultimately be built—the risks of an unknown force stopping the project still exist.

FA is better served by selecting some number of projects, say five, which are ranked at the top. By developing those five in parallel through the increments of project development, the agency can monitor progress and judge risk at each stage, forcing the five projects to compete throughout the development process for the incremental (and increasing) amounts of funding necessary to move the project forward. If, at any time, information becomes available that puts a given project in question, it needs to be considered for abandonment in favor of another. Thus, the five projects for which funding is available will be the best five projects available. This element of competition is a key concept, but it is not effective if not executed in a disciplined way.

Discipline

An undisciplined process will not yield the intended result. The process of project development must include a disciplined approach to decision making regarding what project to invest resources in. In practice, this is a fluid approach that is constantly re-evaluating prior decisions, given new information that has been gained. The most important area to maintain discipline in is the allocation of resources and the willingness to abandon investment in projects if more favorable alternatives exist.

Appendix B. Project Development Framework Categories

The information provided in this appendix is representative of the type of information collected through the project development process, but is not comprehensive. Specific guidance or recommendations for processing this information and making decisions going forward in the project are not provided, but may be in future FEMP publications.

B1. Site

Site is the first element listed because having legal access to a location with appropriate characteristics for a renewable energy power project is essential to start the development process. Examples of this activity would be a desktop Federal Aviation Administration (FAA) analysis and/or a critical issues analysis ("CIA"). For the developer, site control establishes an ability to recover investments made in the development process. Site is a key ingredient to establish project feasibility; without the fundamentals of Site, Resource, and Off-take in place, a project cannot warrant investments required to fully develop Permits, Technology, Team, and Capital.

Pre-Development Stage Site Elements

During this early-stage of project development, one must confirm that there are no known barriers to conveying the land rights required to execute the proposed project to the developer. It is normal to secure contractual rights to the site at or near the conclusion of pre-development stage.

Because of the importance of site control to project feasibility, initial agreements for site control are established in the pre-development stage for developers. During development, these agreements and conditions must be finalized, then managed and monitored to ensure site control rights are maintained and will ultimately convey the necessary rights to use the site. For Federal agencies, it is important to be cognizant of how the privately financed project will rely on these Site elements in early stages, to identify fatal flaws that do not affect the agency, but will affect the ultimate financing.

For projects in the west, Federal agencies should verify whether the proposed site is administered or affected by the Bureau of Land Management (BLM), which has underlying control of most Federal land in the West. Most Federal agency land in the West has been withdrawn from the operation of the public land laws by the BLM on behalf of Federal agencies so that the Federal agency can administer the land for a stated use or purpose for a specific period, which may be renewed. It is vital to confirm the terms of withdrawal for a Federal mission, such as DOD use terms, because these terms dictate the appropriate use of the land. If the proposed use does not match the original mission and purpose of the withdrawal, BLM may assert its authority to control the land use. This issue is especially important if the agency is planning to host a project that will generate more power than it can use. Exporting additional energy than is needed for the agency site may be considered going beyond current mission

needs by BLM, which may deem the export as a commercial purpose.

Development Stage Site Elements

Once project feasibility and conditions for site control are established it is appropriate to continue the full development of the necessary site and legal documentation to convey the rights to use the site.

Investment in the preparation of site information necessary to close financing and start construction occurs during the development stage. Costs include, but are not limited to, preparation and negotiation of legal documents including contracts defining terms for the transfer of real estate rights; documentation of rights of access including easements and/or rights-of-way; assignability of these rights to third parties including financial institutions; responsibilities of all parties with respect to liability, insurance requirements, and indemnification clauses; and technical information such as land surveys and geotechnical studies.

Inter-relationships: How Site Issues Affect Other Project Development Elements

The key elements of the project development framework are not entirely distinct from one another. They typically have multiple overlapping elements. Some examples of interrelationships of the Site element to other framework elements are listed below.

RESOURCE: Site defines the boundaries, context, and conditions under which the renewable resource is collected and converted to useful energy, goods, and services—Site and Resource are very much linked. At what level does the site provide access to the renewable resource? How does the resource stack up within the marketplace that the site will be producing in? Will site characteristics or proximity limit or penalize the site economically? Example: A gravel mining operation neighboring a solar site might alter the atmospheric conditions (dust interferes with solar resource) or increases costs (operations and maintenance [O&M] from additional cleaning and maintenance costs).

OFF-TAKE: Prices for renewable energy vary widely across geographic markets. Access to wholesale energy markets through electric transmission will have a major impact on project economics. Costs to transmit or interconnect to existing infrastructure that is distant or technically difficult may be prohibitive to project economics, limiting the viability of the project.

PERMITTING: Site location and condition can have a significant impact on the ability to obtain permits for projects. Final site selection can be influenced by permitting requirements for different jurisdictions. The presence of sensitive flora or fauna, wetlands, cultural resources, and other environmental resources, will be evaluated in the NEPA process and through other environmental review and permitting actions.

TECHNOLOGY: Site characteristics may naturally be supportive of certain technologies or certain deployment techniques over others. For example, with all other things equal, a sloped site may be ideal for a fixed-axis photovoltaic (PV) system, and sway the investment decision that direction vs. a perfectly flat site that may perform better with single-axis tracking technology. If a developer favors a particular technology, they may favor sites with unique characteristics and vice versa (sites with particular characteristics may get the most value by supporting a particular technology).

TEAM: Particular expertise may be necessary to mitigate site-specific challenges. Wildlife expert services may be necessary for the life of a project if the site is located within a sensitive area, but other factors drive the pursuit of a project. Costs and timing of those costs will come into play and must be considered in the iterative risk evaluation throughout development.

CAPITAL: Financing incentives and programs can be related to geographic areas. These factors may create unique access to capital sources that make sites in that area more or less valuable relative to projects outside the particular geography.

Example Project Questionnaire – For Federal Sites

Ownership and Control (Installation/Base Overview)

❏ Who owns the real estate? Is BLM involved?

❏ Are there conflicting real estate rights held by different parties?

❏ List all parties necessary to legally convey rights, and document the process, if any, necessary to execute a transfer.

❏ How will site control be conveyed? Examples might be an exclusive right-to-build granted for a defined time period such as 90 days, or an option contract that may include periodic payment to maintain exclusive rights over many years.

❏ What are the terms and conditions, including payment, to achieve conditional site control? Have any mission impacts been de-conflicted?

❏ Can the site control be transferred, sold, or assigned to another party?

❏ What rights are necessary for the project? Examples might include a LUA, sale, or easement of use.

❏ Can the granting party agree to a subrogation of rights if necessary to support financing of the project?

❏ Is there any risk of cost recovery imposed by either party with respect to costs incurred?

❏ Are there FAA restrictions that impact the site?

Access

❏ Is the site accessible directly by public roads?

❏ Will existing access accommodate all needs during construction, operations, and maintenance phases? Consider construction equipment, labor force, and specialty equipment such as wind turbine blades and tower elements.

❏ Does site access require crossing or impacting property owned or controlled by another party?

❏ s access controlled by a fence, gate, or security of some kind?

❏ Are there safety, insurance, or liability requirements for project employees, contractors, or agents visiting the site?

❏ What is the procedure to gain access on a regular basis?

❏ Are there conditions under which access would be restricted or eliminated?

❏ In the case of flood, fire, or other natural disaster or emergency, can emergency crews reach the site (including utility crews who may require access and control of the generation equipment)?

❏ Do access easements or rights-of-way need to be established and conveyed?

❏ Are legal permissions and/or physical instruments necessary for ongoing access (to avoid trespass; locks, keys, notification procedures, etc.)?

❏ Which party is paying for costs that may arise to arrange for, maintain, or execute site access?

❏ When will that payment need to be made, does any party have specific requirements on timing?

❏ If the project is not built, is there any risk of cost recovery imposed by either party with respect to costs incurred?

Physical and Political Characteristics (Geography and Land Use Master Planning Data)

❏ How is the site area defined? Examples may include a legal description, land survey, parcel map, roof or structure boundary.

❏ What is the site mission?

❏ What is the potential project impact?

❏ s the definition used sufficient for temporary and long-term site control?

❏ What are the zoning or land use regulations that apply to the site?

❏ Does the proposed use conform to all land use regulations at this time?

❏ Are necessary adjustments, exceptions, or entitlements to land use regulation subject to documented procedures and policies? What are they?

❏ What are the regulating bodies that have a say in land use on the site?

❏ Are neighboring uses likely to conflict with the proposed project?

❏ Do local communities, homeowners associations, neighborhood groups, or other organized groups exist in the vicinity of the project?

❏ What procedures or requirements are imposed by a government or other authority to develop the site?

❏ Do adjacent land uses impact the ability to capture the renewable resource on the site? Examples might include obstructions to wind resource or shading or other obstruction (excessive dust) impacting solar resource collection.

❏ Do adjacent land parcels have the legal right to build or develop structures that would impact the ability to capture the renewable resource on the site? Example might be a possible shading structure on an adjacent parcel, with the owner of that parcel having the legal right to build such a structure in the future.

❏ Are there any airport zones or aviation activities nearby?

❏ What is the topography of the site?

❏ What are the drainage characteristics?

❏ Is there existing vegetation on the site? What is it?

❏ Are existing structures documented?

❏ Do any historic structures exist on the site? Are any known to have existed?

❏ Do any cultural resources exist on the site?

❏ Archeological resources?

❏ Is the buildable area sufficient to support ancillary infrastructure?

❏ Are there any constructability issues to be mitigated? Anything that interferes with standard means and methods for engineering/construction industry?

❏ List all property encumbrances. Does a title policy exist or has title research been performed?

❏ Do any flood zones exist?

❏ How will fire protection be accomplished?

❏ Are there geologic hazards or seismic zones?

❏ Is groundwater present? At what level and seasonality? Will this impact geotechnical requirements?

❏ Has the site been developed at any prior time or is it undisturbed?

❏ Is there an environmental report in existence for the site such as a Phase I or Phase II environmental investigation?

❏ Is there evidence of wildlife activity on or across the site?

❏ What are the geotechnical characteristics of the site, including surface and subsurface soil types? Is Geographic Information System data available?

❏ Is there a geotechnical report for the site?

❏ Which party will incur the cost of generating any necessary documentation for the site?

❏ When is the cost anticipated to occur? Does either party have a requirement for timing?

❏ If the project is not ultimately built, is there any risk of cost recovery imposed by either party with respect to costs incurred?

Technical Integration and Interconnection Information

❏ Potential off-takers

❏ Interconnection points

❏ Distance from project sites

❏ Transmission or distribution

❏ Line ownership

❏ Substation ownership

❏ Line capacity

❏ Interconnection limits

❏ Planned transmission upgrades

❏ Feasibility study

❏ Facility study

❏ System impact study

❏ Utility assessment

❏ Primary electricity provider details

❏ Secondary electricity provider details

❏ State and local utility regulations

Costs and Schedule/Project Milestones/Financial Analysis Inputs/Detail Acquisition Approach

❏ What are the costs and schedule impacts associated with establishing all elements listed above, including

- …impact of obtaining conditional site control?

- …impact of obtaining permanent site control?

- …impact of gaining access rights to the site?

- …impact of obtaining project entitlements (all approvals or permits necessary to have the legal right to build the project)?

- • ...impact of establishing all physical characteristics of the site?

- • ...impact of mitigating, correcting, or altering the site characteristics to suit the project?

❑ Are any cost or schedule requirements out of the ordinary, acting as a burden to the competitiveness of the project?

❑ Do any cost or schedule attributes provide the project with a competitive advantage in the marketplace?

❑ How are cost or schedule advantages or disadvantages accounted for or mitigated? Do they impact the price of power or other attributes contributing to project revenues?

❑ Do cost or schedule advantages or disadvantages impact financing timing or the cost of capital?

❑ Are impacts of cost and schedule passed on to the site owner (through the terms or price of site control), or the off-taker (price of goods or services, or terms of delivery), the developer (profit, risk, experience), or any combination?

For Comprehensive Installation Assessment

See Appendix H. Project Validation Report (DRAFT) for a project validation workbook with detailed outlines on addressing many of these details

B2. Resource

This category is focused on the renewable resource that is the feedstock or raw material to generate renewable energy in the form of heat or electricity. Whether using solar, wind, geothermal, biomass, or other renewable resources as defined on page 2, each is displacing a conventional fuel supply, such as natural gas, oil, or coal, which feeds a traditional energy generating station. Developing an understanding of the quality and quantity of the resource potential at a site begins with a general characterization, done with readily available mapping and data sources, and can end with highly refined engineering data that closely define and predict plant operations and output.

Resource is a key element in qualifying a project and, along with Site and Off-take, is an important first step in identifying a good project. With Site, Resource, and Off-take elements characterized with some degree of confidence, investment in the other elements and deeper investments across the project can be justified.

Pre-Development Stage Resource Elements

Characterization of renewable resources involves investment in the costs of engineering professionals, data collection, and, in the case of a wind project, installation of temporary monitoring facilities for 1 to 2 years to verify the renewable resource for a particular site. In the pre-development stage, when the greatest uncertainty exists, resources are generally characterized using national or regional mapping data (e.g., NREL), publically available data from nearby weather stations or resource monitoring stations, and some limited site investigation. Based on this characterization, the probability of a successful project must be

assessed and the decision made whether to pursue the project over any alternatives that are available.

Development Stage Resource Elements

Investment in resource engineering and analysis, data collection, and modeling is pursued in the development stage. The result of this investment is to increase the confidence factor around the productivity of a particular resource, effectively reducing the expected error in production estimates. As an example, for a solar project, uncertainty of annual or monthly production for a particular site can be estimated to +/- 10% to 20% using desktop studies with minimal investigation—after full characterization using engineering methods, that uncertainty can be reduced to 3%, or a 97% confidence factor.

The tradeoff between the certainty that resource engineering provides and uncertainty is obviously the cost of the analysis, but also the risk of incurring the cost and not achieving the project, thus losing the chance for cost recovery. These decisions must be made, and the burden is typically left to project developers to decide the extent, and timing, of resource engineering. Much of this is influenced by the demands of capital providers to a project; banks and lending institutions seek a high degree of certainty, equity investors or other structured investors may tolerate more uncertainty if able to demand a higher return in compensation.

Inter-relationships: How Resource Issues Affect Other Project Development Elements

The key elements of the project development framework are not entirely distinct from one another. They do relate or typically have multiple overlapping issues that cause many subjects to affect, or be affected by, multiple elements. Some examples of interrelationships of the resource element to other framework elements are listed below.

SITE: Site defines the boundaries, context, and conditions under which the resource is collected and converted to useful energy, goods, and services—Site and Resource are very much linked. At what level does the site provide access to the renewable resource? How does the resource stack up within the marketplace the site will be producing in? Will site characteristics or proximity limit or penalize the site economically? Example: for a solar site, a gravel mining operation neighboring the plant might alter the atmospheric conditions (dust interferes with solar resource) or increase costs (O&M from additional cleaning and maintenance costs). Are there trees that may grow and shade the PV array?

OFF-TAKE: Particular resources may have value to particular off-takers, due to the timing, volume, quality, predictability, or policies within a market. Some market regulators or governing bodies may place value on a particular resource, where others have not. Resource availability must be matched to an off-take arrangement that creates value.

PERMITTING: Land disturbance, subsurface disturbance, project size, or need for additional infrastructure of ancillary facilities can complicate or streamline the necessary permitting processes for a project.

TECHNOLOGY: Resource and Technology are closely linked; resource characteristics can heavily influence the choice of technology or design characteristics of a completed system. A project may move from a "wind project" with a generic placeholder for specific wind turbine technology to establish feasibility in pre-development to the choice of a specific wind turbine design, technology, or vendor that is specifically matched to the resource characteristics.

TEAM: Qualified engineering support, industry expertise, and vendors of data and services all are necessary to determine the level of investment required or prudent for a project at different stages.

CAPITAL: Renewable resources are variable and availability may be entirely up to the natural process; as a result, capital providers require high levels of investment in resource engineering prior to committing capital to the development or construction of a project.

Example Project Questionnaire - Resource

❏ Do any long-term datasets exist on or in the vicinity of the site? Examples can include not just renewable resource monitoring sources that are publically available, but also related weather data that record temperature, wind, moisture, extreme events, or air-quality measures over time.

❏ Will atmospheric conditions affect the project? Are there atmospheric impacts from adjacent land uses?

❏ Do neighboring or adjacent land uses interfere with resource capture?

❏ Given existing entitlements, land use regulations, or other authorizations, might future land uses adjacent or in the vicinity of the site negatively impact resource capture?

❏ Has resource been measured on-site in one or multiple locations?

❏ If a large site, has variation in resource been characterized across the entire area?

❏ Have any monitoring stations been installed? For how long?

❏ Has data been properly scrubbed for errors and confirmed for accuracy by a third party?

❏ Has satellite modeling been conducted or is it available?

❏ Is the vendor or technology provider involved in resource engineering and production estimates?

❏ Does the vendor or technology provider warrant the analysis or work? Does financial capacity exist to back up the warranty?

❏ Have trained, experienced, qualified professionals conducted and documented the resource engineering work?

❏ Have on-site data sets been collected over a 1- or 2-year period?

❏ Is there a confidence factor associated with the resource estimate being established according to industry standards?

❏ Have on-site data records been correlated to long-term satellite or computer generated models?

❏ Is any adjustment or fine-tuning necessary that is technology specific or vendor specific?

❏ For offsite feedstocks, such as biomass, have samples been gathered and tested for use?

❏ Does the project schedule reflect the time necessary for data collection, verification, and engineering?

❏ Will additional third-party verification be necessary to obtain financing?

❏ Are there published industry standards for methods and procedures for resource measurement?

❏ Who will be responsible for paying the cost of resource engineering?

❏ When is it anticipated or expected that this work will be completed? At what scope?

❏ Will the costs incurred be subject to loss if the project is not built? Are any parties expecting reimbursement of some, or all, incurred costs?

B3. Off-take

The Off-take category represents all things necessary to achieve a long-term contract with a customer to purchase the output of a renewable energy project. The long term contract(s), or off-take contract(s), establish the revenue profile for the project, which forms the basis of financing and thus heavily influences the feasibility that a project can be constructed.

Prominent in this discussion is the PPA, the most common form of "off-take agreement" with an "off-taker," or purchaser. The subject is more complex than just the economics of a purchase contract in that it also implies or requires the physical ability to get electricity or services to that willing purchaser (over existing or proposed infrastructure including substations, and distribution and transmission lines). The economic ability to do this can also come into play; in other words, many things are physically possible but not economical. Renewable power plants can produce revenue from more than selling electricity; selling heat, ancillary services, and RECs are some examples of other revenue streams.

For a very large renewable energy project on Federal lands with minimal energy needs, the utility is the obvious off-taker of the power. However, for a large Federal facility, such as a military base, the government may be an off-taker of all or of a large portion of the power. Keys to a successful off-take agreement in a utility-scale project are the state laws that govern electricity and energy and whether a state has an RPS or similar requirement for renewable energy. The RPS generally includes some incentives for renewable energy. These regulatory requirements will drive whether an off-take agreement is legal, and whether a utility is interested in buying the energy, and therefore should be reviewed at the earliest stages when considering projects. When the system is sized below the energy use of the site, the off-take agreement will likely only be between the Federal agency and the project owner.

Pre-Development Stage Off-take Elements

It is essential to identify and to qualify the revenue opportunity and characteristics for a project at the earliest stage. For any project, except perhaps isolated, off-grid applications, the local utility and/or local regulatory bodies are essential stakeholders that must be engaged. For larger projects, a utility is likely to be the purchaser of plant output and in that case will be the off-taker. In a regulated market, this may be the local utility that has a service area that covers the site, or it may be a more distant utility connected to the site through an Independent System Operator (ISO) in an unregulated marketplace.

After defining the potential customers, it is important to establish the market price, tenure, and location at which the power transaction is measured and affected. Price is defined as payment for goods or services, and typically varies over the life of the contract, which, depending on the method of financing and parties involved, can range up to 30 years in length. Price is defined for the entire contract period, and its importance is obvious. Tenure refers to the length of the off-take agreement or other contracts that provide revenue sources. It is possible that the market for a particular aspect of plant output, such as RECs, may be for a limited time period of three to five years. The uncertainty after that time period is speculative and introduces risk to the project. Collectively, the tenure, or length in time, of all revenue contracts together define the certainty of revenues and influence overall project valuation.

Location of the purchase/sale exchange is important because it begins to define the more complex issues of delivery to market, access to markets, interconnection processes, and transmission congestion. A high-level discussion is provided in Appendix C. Overview of Electricity Markets and Key Terms as introductory background on these issues.

It is not necessary to have a signed off-take agreement at this stage of development, but it is necessary to establish with reasonable certainty (1) the likely price for electricity, RECs, or any other revenue streams, (2) the length of the contract, (3) the ability to participate in the market, and (4) that the contract(s) are financeable. Market participation includes the ability to complete an interconnection agreement, gain access to the transmission system, and/or build any necessary infrastructure necessary to physically connect the project to the system. This whole set of factors must also be judged to be economical; if for example, the off-take agreements are financeable, but at a rate of interest that is cost prohibitive, the project cannot be completed and should not be pursued. This analysis of economics is accomplished using a financial model called a pro forma. An example pro-forma is provided in Appendix F. Project Pro Forma Example.

Development Stage Off-take Elements

If not achieved in pre-development, an off-take agreement should be established as the first course of action in the development phase, or the project should likely be abandoned or delayed until an off-take agreement can be established. In addition to the off-take agreement, all other agreements or actions necessary to activate the off-take agreement must be accomplished; this refers specifically to interconnection and transmission agreements, each of which will have its own impact to project cost and schedule.

For instance, if transmission access is necessary to get the product physically to the market it is being sold to, it can be a multi-year process to get into the transmission queue and secure access to transmission services.

Professional developers are highly experienced in the energy markets, have access to extensive market data, and maintain relationships with customers. Appendix C. Overview of Electricity Markets and Key Terms provides more details on the electricity market and key terms.

Armed with this information, developers are called upon to make good judgments regarding the opportunity and likelihood that a project will get a financeable off-take agreement. This is being monitored constantly along with other parallel investments necessary to move the project closer to construction and operation, such as permitting processes, technology selection, engineering design, and sourcing necessary capital partners to finance the project. With low levels of confidence, a developer may choose to put off investment in these other areas to mitigate the risk that those investments will be lost if an off-take agreement is not put in place. With high levels of confidence, he or she may choose to invest aggressively.

Inter-relationships: How Off-take Issues Affect Other Project Development Elements

The key elements of the project development framework are not entirely distinct from one another. They do relate or typically have multiple overlapping issues that cause many subjects to affect, or be affected by, multiple elements. Some examples of inter-relationships of the off-take element to other framework elements are listed below.

SITE: Site influences the project's ability to sell its output, the cost of getting the output to customers, and the price paid for it. Siting of projects is executed with the elements of off-take in mind; the electrical system's technical characteristics will also influence the value and viability of off-take agreements because the location of the power is important to its value.

RESOURCE: Renewable resources have different attributes and characteristics in the amount of power production the predictability of production over time, the time of day energy is produced, etc. All of these factors will affect the price paid for the project's output and the terms and conditions that are included. Solar energy is valued in certain markets because it is produced at times of high peak demand on the system; other markets value solar due to policies that support solar power attributes. Wind may be more valuable in large, interconnected markets, but not small, isolated grids, because of concerns over system variability.

PERMITTING: Permitting and Off-take are not as directly related as some other elements. Off-takers and developers are concerned with the risks associated with executing a contract that relies on permits not yet in hand. To mitigate this, timing limitations, conditional approvals, achievement milestones, or other mechanisms may be inserted into permitting or off-take agreements and documentation. These issues must be closely managed and monitored for conflicts that create cost or schedule impacts.

TECHNOLOGY: Sophisticated off-takers, or purchasers of renewable energy, recognize the attributes of the output from different technologies can vary. Even balance-of-system designs can affect output. For large-scale projects in particular, the technology is being relied upon by all parties to perform reliably over extended time periods. Off-take agreements commonly include performance requirements that are key contract terms; if those terms are not met, the contract could be voided or terminated.

TEAM: Marketing, negotiating, and consummating a long-term off-take agreement or other revenue agreement can take specialized skills from multiple disciplines working together. Power marketers or sales professionals, attorneys, engineers, and business managers are some examples of the kinds of experienced team members necessary to accomplish the signing and activation of revenue contracts.

CAPITAL: Project revenue is defined within the off-take category, and revenue is a key driver of a project's ability to provide returns on capital invested and the perceived risks associated with those returns. The credit quality of the purchaser is of keen interest to capital providers, as a contract with an insolvent customer is worth very little and creates tremendous risks for investors. Terms and conditions associated with off-take contracts are also influential in financing decisions; off-take agreements will be scrutinized in great detail by investors.

Example Project Questionnaire - Off-take

❏ What Contracting Authority will be used?

❏ What products and services will be produced and sold to produce revenue for the project?

❏ Does the state have a Renewable Portfolio Standard?

❏ Is the regulated utility interested in buying the power?

❏ Is there more than one perspective buyer for any given element of project output? Who are they?

❏ Do standard contract forms exist, or will they be developed for this project? Who will pay the cost of contract development and negotiation?

❏ Is this a new or established market?

❏ What other market participants, or competitors, exist?

❏ Are there open bidding mechanisms, competitive bidding, or individually negotiated contracts for off-take agreements, PPAs or other revenue opportunities?

❏ Will time-of-day be a factor?

❏ What are the terms and conditions of the off-take agreement or contract?

❏ Can the contract be cancelled or ended by the other party? Is there recourse for the project if this occurs? How is recourse affected? Will legal action be necessary?

❏ Who will market, negotiate, document, and sell the project's output? How will this cost be accounted and paid for?

❏ Is this project expected to be competitively advantaged or disadvantaged in the marketplace?

❏ Is there an application process or formal structure to be granted entry to a market?

❏ Are there policy mechanisms that must be relied upon to have access to the market? What are they? Are they certain and objective, or subjective?

❏ What is the value, or likely price, the project will be paid for its output?

❏ What factors influence the price(s)?

❏ Are the market and price(s) stable or are they highly variable?

❏ Will purchasers sign a long-term contract? What length of time is typical in the market?

❏ What is the location of purchase? Will it be at the project site, at a nearby substation, a distant location, or other?

❏ How will the volume of output be measured and verified? Who pays for this?

❏ Will system output be audited? Are there mechanisms to solve issues with output or disagreements between parties?

❏ Will contracts expire if not activated after some period of time?

❏ Is the electricity market regulated or unregulated?

❏ Are the purchasers credit-worthy and financeable? Would a bank lend money against that customer's commitment and ability to pay for services rendered or goods received?

❏ How long will it take to negotiate and secure an off-take agreement?

❏ Who approves the off-take agreement? Is it the customer? Does a regulator have to approve it?

❏ What steps or procedures must be taken to have access to the market or customer? Are these administrative or selective procedures?

❏ Will a system impact study be required by the utility or load serving entity?

❏ Does electrical infrastructure (substations, distribution lines, or transmission lines) exist on or near the project site?

❏ Does infrastructure being proposed or constructed by an unrelated third party enable this project? Does the project rely on this third party's progress and cooperation? Have both been secured?

❏ Are fees or charges necessary to apply for transmission, interconnection, or off-take agreement?

❏ Does the infrastructure available serve the needs of the project or will it need to be upgraded or changed?

❏ Will new infrastructure (substations, distribution, or transmission lines) need to be built? On the project site or off (same landowner or different)?

❏ Does existing infrastructure have capacity?

❏ Is there a system operator (SO) involved in the process of securing necessary agreements? What are the costs, standards, procedures of the SO?

❏ Is existing infrastructure old or past its useful life? Will the project require that it be replaced? Will the project bear a cost burden?

❏ What will happen at the end of the off-take agreement or other contract term? Will plant output still be viable and sellable? Will follow-on contracts be available at prevailing market rates?

❏ Is the value of future contracts accounted for in the financial analysis of the project?

❏ Who will pay the costs of any work or direct cost necessary to accomplish all agreements necessary to affect off-take agreement or revenue contracts?

❏ When will these costs be incurred?

❏ What if the project is not built? Are any parties to off-take agreements or other revenue contracts expecting reimbursement of some, or all, incurred costs?

Baseline Energy Data

❏ Megawatt hours/year

❏ British thermal units/year

❏ Average demand

❏ Peak demand

❏ Pertinent additional usage information

❏ Existing renewable energy projects

❏ Renewable energy project power inputs

❏ Energy requirements assessment

❏ Current energy security measures

❏ Current meter locations

❏ Planned meter locations

Technical Integration and Interconnection Information

❏ Potential off-takers

❏ Interconnection points

❏ Distance from project sites

❏ Transmission or distribution

❏ Line ownership

❏ Substation ownership

❏ Line capacity

❏ Interconnection limits

❏ Planned transmission upgrades

❏ Feasibility study

❏ Facility study

❏ System impact study

❏ Utility assessment

❏ Primary electricity provider details

❏ Secondary electricity provider details

❏ State and local utility regulations

B4. Permits including NEPA Compliance and Permitting Activities

Projects may require a variety of permitting before construction can be started; this category identifies a number of potential approvals, permitting actions, or processes that, if not achieved, may stop the project. The category includes everything from local building permits and internal authorizations to satisfaction of NEPA requirements. With some limited exceptions, Federal agencies must comply with NEPA before they make final decisions about proposed actions that could have environmental effects. This discussion is not intended to be all inclusive. Agencies should check with their appropriate counsel.

Pre-Development Stage Permit Elements

Permitting can be a time-consuming process and is resource intensive in terms of human resources and money. In the earliest stages, identifying all necessary permitting activities and documenting the requirements is the first step. This is done while looking for fatal flaws, or the expectation of serious conflicts that may challenge the ability of the project to be permitted.

While no permits are typically secured at this point in the project, coordination with authorities should occur to establish the cost and schedule impacts of each permit process, along with the information required to proceed. As an example, acquiring building permits requires project plans, engineering drawings, and specifications. The timing of the permitting process will be paced by the development of this level of project documentation and the willingness of the project team to expend the resources on design teams.

Any authorities having jurisdiction would be involved in early planning, and all applicable codes and standards would have to be observed. A number of organizations promulgate standards related to large renewable energy systems interconnected with the utility system. These organizations generally address performance, safety, and communications. Standards that apply to only "utility interactive" systems are different than those that allow a system to provide power to a microgrid when the utility grid is down, and a renewable energy generating system would have to accommodate both if it were to contribute to energy security.

Development Stage Permit Elements

The major investment in permitting is executed in the development phase. At this point in time, the project should have passed several iterations of fatal flaw analysis and has matured with regard to design to the point that it can be clearly explained to permitting authorities and stakeholders. For projects executed outside of the Federal context, the private development community has expertise on the permitting process, managing the investment and progression of it, and mitigating the risks involved.

In the Federal context, the NEPA compliance process is added to the requirements of permitting and entitlements necessary to build a project. All of these requirements may collectively be referred to in the development community as "permitting risk." The NEPA compliance process adds time, uncertainty, and development expense to the development of a project. Private developers may or may not be willing or able to carry the costs of NEPA compliance work and at the same time be comfortable with unknowns that the process could introduce in terms of project design and function. Further, the time taken for compliance introduces the opportunity for other project inputs to change; project direct costs, financing costs, revenues, or policy changes are some examples of key parameters that can shift over time. All of this suggests that perceived risks from individual project circumstances may impact the private sector's ability to finance NEPA compliance as a development activity. A Federal agency's ability or willingness to support the cost of NEPA compliance is not, however, a complete solution because decisions made through NEPA compliance have direct technical and financial impacts to project feasibility, impacts that are best informed by expertise from the developer. These issues must be balanced and will be decided based on individual project characteristics. NEPA

compliance is further discussed in Chapter 3 of this document in both the Stage 2; Validation and Stage 3; Acquisition sections.

Inter-relationships: How Permit Issues Affect Other Project Development Elements

The key elements of the project development framework are not entirely distinct from one another. They do relate or typically have multiple overlapping issues that cause many subjects to affect, or be affected by, multiple elements. Some examples of interrelationships to other framework elements are listed below.

SITE: Site influences the project's permitting requirements because it establishes not only jurisdictional controls, but also specific elements that may trigger permitting, such as the existence of sensitive environmental conditions. Two sites that are otherwise equal may be distinguished in priority due to differing requirements or perceived risks.

RESOURCE: The methods of conversion for different renewable resources may have an impact on permit requirements. For example, geothermal projects have a different set of potential environmental impacts (such as groundwater and surface disturbance patterns) than large solar PV or wind projects.

TECHNOLOGY: Technology decisions can also impact permitting; a strong solar resource could be collected and converted to electricity by either PV or concentrating solar power (CSP) technology. Some CSP technologies require significant water resources to operate; PV does not. Permits for the use of significant water supplies may interfere with the selection of CSP.

TEAM: Permitting is a specialization that requires professionals who practice, at times exclusively, in this area of work. This must be considered when assembling the project team.

CAPITAL: The investment community will resist making investments until they are sure the project can be built. Overall permitting risk, including compliance with NEPA, is an important risk to be considered by an investor when projects are competing for investor's capital.

Example Project Questionnaire - Permits

❏ What permits or authorizations are required from the local utility, regulating body, or associated stakeholder?

❏ What permits or authorizations are required within the agency or organization that owns the site?

❏ What permits or authorizations are required from local jurisdictions or agencies?

❏ What permits or authorizations are required from state jurisdictions or agencies?

❏ What permits or authorizations are required from Federal jurisdictions or agencies?

❏ What permits or authorizations are required from non-government entities?

❏ What information does each permit or authorization require in order to be processed?

❏ Is that information available or will cost be incurred?

❏ What is the cost, and when must it be expended to meet the schedule of the project?

❏ Is litigation of permitting issues expected or probable?

❏ Could the timing of permits and authorizations significantly impact the costs or economics of the project? Could that put the project in jeopardy? (Is the project time-sensitive in a key area, such as revenue contracts, that may expire prior to achievement of all authorizations and permits?)

❏ Who will pay the costs of any work or direct cost necessary to accomplish work necessary to achieve approvals, authorizations, or permits?

❏ When will these costs be incurred?

❏ What if the project is not built? Are any parties in the process expecting reimbursement of some, or all, incurred costs?

B5. Technology

The technology category addresses the selected technology for a project, and the process to design and engineer all necessary facilities. The design process (and associated investment) is embedded here. As an example, it may have been determined that a project is expected to use PV technology; this is the "prime mover" or key conversion technology. Beyond the selection of a PV panel vendor or supplier, the project will require significant design effort to select and engineer components for everything from foundation structures and electrical connections to inverters and transformers.

Most large RE systems are designed to feed power only to an operating utility system. Special configurations of hardware and operating sequences are required for the RE system to contribute to energy security. This capability would also affect the communications and control with the utility. Such measures add to the cost of the system, yet do not contribute to a revenue stream derived entirely from the sale of bulk kWh to the facility or utility.

Pre-Development Stage Technology Elements

At the earliest stages, and perhaps through the pre-development stage, technology selection is not intensive or highly detailed. Technology is a category that is generally more under the project developer's control and influence than many others, such as permits or off-take—both of which depend on the cooperation and agreement of outside parties. Because of this, early investment may be placed in those categories, with a generic technology assumption used as a placeholder in technical and financial analyses.

Development Stage Technology Elements

Once a project progresses to the development stage, major investment is going into all aspects, including technology. In this stage, initial assumptions and/or conceptual (35% complete) engineering drawings are brought through the entire design phase including design development (60% complete) and, ultimately, construction documents (100% complete). Depending on the organization of the project team, a third-party firm may be providing engineering, procurement, and construction services (EPC contractor), or the developer may hire independent engineers who will use the plan sets for bidding to construction contractors.

Inter-relationships: How Technology Issues Affect Other Project Development Elements

The key elements of the project development framework are not entirely distinct from one another. They do relate or typically have multiple overlapping issues that cause many subjects to affect, or be affected by, multiple elements. Some examples of interrelationships to other framework elements are listed below.

SITE: Site characteristics influence technology through selection of preferred components to specific project design elements such as foundation systems or structural systems. The site must be well understood and characterized for engineers to be able to prepare accurate design plans and make design decisions along the way.

RESOURCE: Technology and Resource categories are inexorably linked, as the conversion technology and associated systems are likely selected and tuned to take advantage of the site-specific resource characteristics that are provided.

OFF-TAKE: The off-taker of the electrical power, heat, or services from a renewable project may have to meet system design and operational standards to interconnect (in the case of the utility), or tie into building systems—all can be related to the development of the system design and may become known in the PPA or off-take contract.

TEAM: Designers and engineers are key team members who must be engaged along the way to accomplish the needs of detailed drawings and specifications for the project.

CAPITAL: Investors and lenders are interested in minimizing project risks; technology includes the key operating systems that produce products and ultimately, revenue. They will have keen interest in the reliability and track record of selected technologies and systems.

Example Project Questions—Technology

❏ What is the technology?

❏ Does the technology have a record of successful commercial operation?

❏ What risks exist that would threaten plant output due to technology failure?

❏ How much will it cost to purchase, install, and commission?

❑ Does the technology chosen meet national certification standards (e.g. ANSI standards)?

❑ Will incremental payments be necessary?

❑ Is there a time requirement for delivery between placing an order and receiving the equipment for this technology?

❑ Will incremental payments be required by the manufacturer?

❑ Is the technology designed and selected to perform in this particular environment? Could it be susceptible to problems in certain environments (excessive heat, wind, moisture, salt water, etc.)?

❑ Will performance change over time? What factors will influence degradation of performance?

❑ Are performance warranties and guarantees available from manufacturers or vendors?

❑ Does the provider or warranties or guarantees have the financial capacity to perform?

❑ Is there a detailed operation and maintenance plan developed for the plant?

❑ Is it documented and fully funded for the life of the project?

❑ Will the technology require replacement parts or periodic capital investment?

❑ Does local expertise exist to install, operate, and maintain the project's equipment?

❑ Are service contracts in place for the life of the project?

❑ Insurance will be necessary—are those costs included?

❑ Are designers and engineers licensed?

❑ Do they carry professional liability insurance?

❑ Is the design and engineering team experienced, with a track record of successful projects?

❑ Who will pay the costs of any work or direct cost necessary to acquire technology selection analysis, design, engineering, or other work on offsite facilities?

❑ When will these costs be incurred?

❑ What if the project is not built? Are any parties in the process expecting reimbursement of some, or all, incurred costs?

B6. Team

Every project requires a project team to execute. The expertise of many professionals will be required at different points in time, some for the length of the project. Engineers and architects, attorneys, financial advisors and modelers, accountants, sales and marketing professionals, business managers, negotiators/lead

project officers, environmental and permitting specialists, etc. can all be necessary.

No one entity provides the entire team with all the expertise needed; in the broadest application of this category, each stakeholder who has a key role in the project's success is viewed as a team member. As an example, a Federal agency that is leasing its site to a private developer who is in turn selling renewable power to the local utility requires teamwork from all three key participants—agency, developer, and utility—to be successful.

Pre-Development Stage Team Elements

Because the outcome is uncertain, the earliest phase of the project is performed with a very small team to limit the cost impact to the project. A team of one to three people would likely be able to assess and screen multiple early-stage projects; one effort in this phase is the identification of team members who will be needed through the life of the project, or at a minimum through its construction and commissioning.

For the government, the team is a critical element of obtaining stakeholder consensus at the concept and acquisition approval milestones. It is vital to ensure comprehensive representation of government stakeholders in the pre-development stage.

Development Stage Team Elements

The development phase involves the greatest level of teamwork as the project engages a full complement of professional services and stakeholder members to step through the iterations of development and be prepared for execution.

Inter-relationships: How Team Issues Affect Other Project Development Elements

The key elements of the project development framework are not entirely distinct from one another. They do relate or typically have multiple overlapping issues that cause many subjects to affect, or be affected by, multiple elements. Some examples of interrelationships to other framework elements are listed below.

SITE: Site selection and inspection visits notwithstanding, the categories of Team and Site have minimal interrelation, save for any specialists a particular site selection may require later in the process.

RESOURCE: Professionals may be engaged for resource assessment.

OFF-TAKE: The identification and negotiation of an off-take agreement requires significant engagement of team members, and may influence the selection of specialists or professionals with key experience.

PERMITS: Permits and authorizations require specialized support that may influence the organization or membership of a project team.

TECHNOLOGY: Design professionals and engineers will be essential to developing a strong Technology category.

CAPITAL: Early-stage investors are likely to act as key team members in the identification and mitigation of risks, as well as the decision to allocate additional capital along the way. For the government team, these roles are taken by those who can allocate staff or funds to conduct early-stage analysis of key elements and risks. Later-stage investors and lenders will view the team element with a discerning eye, seeking a comprehensive team of professionals with technical depth and experience to mitigate unforeseen risks.

Example Project Questionnaire - Team

- ❏ Has the team worked together before?

- ❏ Are there an experienced project manager and subject matter experts working for both the government and developer?

- ❏ Are all stakeholders represented?

- ❏ Who is/are the decision makers?

- ❏ Who is the contracting officer?

- ❏ Does each team member have experience with similar projects?

- ❏ Are all members insured with professional liability insurance where possible?

- ❏ Who will pay the costs of assembling and managing the team?

- ❏ When will these costs be incurred?

- ❏ What if the project is not built? Are any parties in the process expecting reimbursement of some, or all, incurred costs?

B7. Capital

The financial resources required to pay all costs necessary to build a project can likely be attracted to the project once all categories of development are fully developed and unknowns eliminated. The resources necessary to get to that stage are not the same as those before financial close; these development risk capital investments are typically recovered at the point of project financing and start of construction.

Capital resources are typically a mix of debt and equity providers, including tax equity investors, banks or institutional lenders, and other grants or government support for renewable energy projects. Many more complex sources exist, such as vendor-financing or government or corporate bond financing, but the message is the same throughout—for the elements of project finance, a rigorous project development process will have to already have occurred and be fully documented in order to attract capital.

Pre-Development Stage Capital Elements

The ultimate destination of a project development effort is to enable construction to commence and get to an operating, revenue producing project. In the earliest stages, the project must be judged with this destination in mind. Financing sources will require certain kinds of information, risk mitigation vehicles, and rates of return. Being cognizant of the level of each that the capital markets require is necessary to judge the viability of a project and the investment necessary along t e needs of capital providers can lead to imperfect and potentially very expensive mistakes.

Development Stage Capital Elements

Throughout the development stage, a professional judgment is required of the same factors as noted in pre-development, but with higher degrees of accuracy and alignment with market conditions. As larger sums of time and money are invested, an eye to the ultimate destination of financing is important to make informed decisions on incremental investment, acceleration of investment, or the decision to delay or abandon the project.

In the later part of the development stage, the project investment opportunity is brought to the capital markets to be funded by the different tiers of participants (debt, equity, or other). Contracts for Engineering, Procurement, and Construction (EPC) are negotiated and prepared to document the investments, all categories of information in the development framework are reviewed for accuracy and completeness, and finally these agreements are "closed," whereby the commitment to fund is final (subject to the terms, conditions, and covenants).

Inter-relationships: How Capital Items Affect Other Project Development Elements

The key elements of the project development framework are not entirely distinct from one another. They do relate or typically have multiple overlapping issues that cause many subjects to affect, or be affected by, multiple elements. Some examples of interrelationships to other framework elements are listed below.

SITE: Capital investors will examine site selection, control, and rights of access as part of their investment decision.

RESOURCE: Capital investors will examine resource engineering and performance or output predictions as part of their investment decision.

OFF-TAKE: Capital investors will examine the terms, conditions, and economics of the off-take agreement very carefully as part of their investment decision. Many investors or lenders will fully underwrite, or confirm, the off-taker's financial capacity to pay along with the terms of the contract—the credit of the off-taker will influence investment and lending decisions.

PERMITS: Permits must be in-hand and proven to be valid at the time of financing.

TECHNOLOGY: Capital providers will require documentation of the technology system's performance and any warranties and/ or guarantees provided to ensure the technology will perform as advertised for the life of the project.

Example Project Questionnaire - Capital

- ❏ Are the funds available and fully authorized for investment?

- ❏ What is the decision process to approve investment or lending?

- ❏ What information is required prior to investment or lending approval?

- ❏ What requirements must be met to issue future payments?

- ❏ Will funds be available immediately, or drawn over time with progress?

- ❏ Will the lender require inspections and approvals of incremental draws?

- ❏ What are the timing differences of each capital provider?

- ❏ Is there any potential mismatch between timing of funding that must be carried by the project?

- ❏ What is the cost of capital?

- ❏ How will capital be returned?

- ❏ Will priority repayment be made to the lender?

- ❏ Have all investors agreed to subordinate their returns to others, as required?

- ❏ Are there fees or closing costs?

- ❏ Will legal documents be needed? Are they standard, or will they be customized?

- ❏ Has the lender or investor participated in a project of this nature before?

- ❏ What liabilities are created between parties?

- ❏ Who will pay the costs of assembling, negotiating, and administering funds?

- ❏ When will these costs be incurred?

- ❏ What if the project is not built? Are any parties in the process expecting reimbursement of some, or all, incurred costs?

Appendix C. Overview of Electricity Markets and Key Terms

Any power plant that generates electricity will do so in a market context; electricity is a valuable commodity that is generated, bought, and sold through a variety of market structures in the United States. It is important to have some understanding of these markets and recognize that any generator, whether operating from traditional fossil-based fuels, hydropower, nuclear, or renewable energy sources, will be doing so within the constraints and competitive environment of one or more energy markets. In addition to competition between generators, multiple markets and sub-markets may come into play, including the electricity transmission market, ancillary services markets, wholesale and retail markets, capacity markets, and many more.

Given the size and complexity of the system that services an estimated 100 million individual customer meters, any analysis exercise requires an integrated understanding of the industry, including technical, economic, regulatory, and business factors. The intent here is to highlight some significant elements and encourage new participants in this field to seek out more information on the subject, not fully detail all attributes or even all parts of the market and its many participants.

Utility Market Structures

Utilities dominate the electrical energy markets in the United States, and various forms exist including investor-owned utilities (IOUs), cooperatives (COOPs), municipal-owned utilities (municipals), utility companies focused on generation, those focused on transmission and distribution, and those focused on serving retail users or "load" on the system. In almost all cases, there is a wholesale level of pricing and a retail level; much like any vertically integrated industry, those tiers of pricing are at times controlled by a single company and at other times by competitive, open markets at both levels (and the servicing infrastructure in between). The Federal Power Marketing Administrations (WAPA, BPA, SWPA), with the special authority to market power generated on Federal land, have been instrumental participants in many projects within their service areas.

In the United States, important definitions about utility markets are determined by state laws. Key elements are whether the state energy markets are regulated or deregulated. For renewable projects, the state's RPS, which defines state renewable goals, and possible financial incentives are also very important. This is in addition to specific Federal programs, for example the investment tax credit (ITC) program and others, which are assumed to be used by private developers in most every case.

A regulated market is a closed market meaning it is closed to competition. The electricity provider is often a full-service utility that generates, transmits and distributes electricity in a defined service area to retail customers. The company has a monopoly in the service territory and other providers may not enter the market. In exchange for this monopoly, the company is statutorily subject to rate regulation by the state. The state utility commission generally sets rates to allow the utility to collect its necessary and reasonable costs plus a reasonable return on its investment. In addition, the monopoly utility is usually required to provide adequate and reliable service to all customers in its service territory.

At the wholesale level, the generation market may be open, which creates a wholesale market for electricity inserted into the transmission system. The Federal Energy Regulatory Commission (FERC) regulates the transmission and sales of inter-state wholesale electricity. In areas where there are regional transmission organizations or independent system operators (RTO/ISOs), transmission services for the movement of electricity across the grid from generation to load are provided under an open access tariff, administered by the RTO/ISO, and regulated by FERC. In areas where there is no RTO/ISO, the transmission service self-provided by the utility or transacted under bilateral contracts are still subject to the FERC open access requirement and Federal regulation.

It is essential to have some understanding of the market context that private investors and developers will be operating in for a particular project effort.

Balancing Authority Areas

Electrical systems must be balanced, meaning that to maintain equilibrium at any given moment in time the total power generated or available to the system must equal the power being consumed down to the micro second. This is no simple feat, and can never be accomplished perfectly; luckily, the system can withstand minor mismatches between supply and load. It does so by sacrificing power quality (discussed in ancillary services section below), but there are limits to the extent this can happen before the system malfunctions.

To maintain order, Balancing Authority Areas (BAAs) have been formed. These are essentially geographic service areas within which balance is maintained and controlled first by the operating assets and service providers within the area, and second through interconnection to neighboring BAAs. Balancing Authorities (BAs) are organizations that oversee the operations within a given area. BAs can direct the import or export of energy to maintain system balance. A collection of interconnected BAAs is overseen by another level of reliability coordinator, who is responsible for the reliable operation of the bulk power system (generation and transmission) among the interconnected control areas. The Western Electricity Coordinating Council is an example of a reliability coordinator for what is known as the "Western Interconnection," which covers much of the western U.S. There are more than 100 control areas and 10 reliability coordinators in North America. A more detailed description of the functional model used in the United States for the bulk electric system can be found here: http://www.nerc.com/files/Functional_Model_V5_Final_2009Dec1.pdf.

Transmission and System Operators

The ability to move electricity and power from where it is produced to where it is needed, at the time it is needed, and at the level it is needed requires a complex interaction of infrastructure and services to support the system. Transmission lines that provide pathways for electricity to move very long distances at high voltages are an essential element, and the cost to construct and maintain these facilities is financed and recovered within the context of the various market structures previously discussed.

In RTO/ISO areas, the transmission-owning utilities cede control of their transmission facilities to the RTO/ISO and the coordination of these separately owned and financed transmission providers is done by an RTO/ISO.

Congestion

Transmission lines are designed and built at some level of capacity, and have limits to what levels of power or electricity they can carry. They are all interconnected so each transmission line added to a system affects the performance and load of every other component, or line, of that system. "Transfer Capability" is the measure of the ability of the transmission system to move or transfer power (megawatts) in a reliable manner from one area to another over all the interconnected transmission lines (or paths) between those areas under specified system conditions.

Congestion occurs when a particular line is at capacity and it limits the ability of other lines to carry extra load to serve demand. This means that given a particular pattern of load or demand on the system, the opportunity to insert generation at any given point (geographically) can be limited by congestion on the transmission system, and may not be alleviated simply, cheaply, or within a reasonable period of time. In other words, congestion can limit development of new generation.

Ancillary Services

The laws of markets and economics are important, but electricity follows the laws of physics first and foremost. Power quality is a term sometimes used to refer to the characteristics of electricity. Power quality used here is a generic term to refer to a consistent set of characteristics for delivered electricity, within tight performance tolerances. Essentially, industry standards define these characteristics, which can vary in many technical ways, different voltages, frequency, current, etc.

To maintain power quality and consistently deliver electricity within the tolerances of the product specifications requires ancillary services. These are services provided to the overall electrical system to maintain power quality as system load changes in fractions of a second. In RTO/ISO markets, these services are procured competitively and paid for by service providers in the market. In non-RTO/ISO markets, this function is being performed by facilities within the utility's system, and the cost is typically recovered through the margin between wholesale and retail rates.

The technology in this market continues to develop, and regulators continue to emphasize the importance of energy storage in the use of intermittent, renewable energy during times of high demand. It is understood that the integration of energy storage could completely alter the scope of any project, but the potential for the technology to benefit a project's economics may make it a valid consideration.

Renewable Markets and RECs

Electricity generated by renewable sources can be demanded within a market because it fundamentally outperforms other sources economically. Market policies also create demand for renewable energy. RPSs are established in many states and require that a specified percentage of total energy consumed in a jurisdiction be generated by renewable energy. This creates demand, or an appetite, by market participants, such as utilities, to purchase energy from renewable sources in order to meet these requirements.

Markets differ across the country, mostly varying by policies at the state level—the details of those policies can significantly influence the value of a unit of renewable energy as well. RPS markets are very competitive and can be largely met (on paper) by proposed projects, over-subscribed with bids to meet this demand, or underserved by lack of suppliers. The bottom line is that the laws of supply and demand are a factor in these markets.

Because of the limitations of the electrical system and markets discussed herein, renewable energy projects cannot be built just anywhere and at times cannot be approved at all by the local utility or others with regulatory authority. In order to provide flexibility to utilities and others with demand for energy from renewable sources, renewable energy certificates (RECs) were created. RECs document the attributes of a unit of electricity to certify that it was generated from renewable sources. In some markets, RECs for a unit of energy can be bought and sold separately from the energy itself, creating a separate source of revenue that can be used to finance projects.

National Organizations

FERC is an independent Federal agency that regulates wholesale sales of electricity and transmission of electricity in interstate commerce. FERC also regulates the interstate transmission of natural gas and oil, and regulates natural gas and hydropower projects. More information on FERC can be found at www.ferc.gov.

The North American Electric Reliability Corporation (NERC) ensures the reliability of the North American bulk power system. NERC is certified by FERC to establish and enforce reliability standards for the bulk-power system. NERC develops and enforces reliability standards; assesses adequacy annually and monitors the bulk power system, in addition to other activities. More information on NERC can be found at www.nerc.com.

Both FERC and NERC play different, but prominent, roles in setting and enforcing standards, policies, strategic direction, and many other functions for the industry; being familiar with their presence and roles is helpful and another resource for market participants.

Appendix D. Commercial Project Financing

Understanding the basics of commercial project financing is essential for any Federal employee involved in renewable energy project development. Many readers will understand the principles of home mortgage financing; however, an entirely new language appears when the large sums of money required to finance large-scale renewable energy projects are considered. The cost of financing cannot be so large that it acts as a barrier to getting the transaction done, but it must be large enough to satisfy investor's required returns.

> The cost of financing cannot be so large it restrains the transaction, but it must be large enough to satisfy investor's required returns.

This section will provide some background that may demystify the "art" of project financing, helping with the translation between the government and the private sector. There is an important added benefit in recognizing that the translation goes in both directions; the private sector also has a need to understand the language of the Federal Government.

What Makes a Deal?

Regardless of the market sector, large-scale renewable energy projects include the integrated efforts and commitment of multiple parties, suppliers, and service providers, all of which operate subject to market dynamics. One of the most common reasons a deal fails is that a mutual understanding of the goals and constraints of each party was not established upfront, resulting in wasted time and lost development funds.

> One of the most common reasons that deals fail is that a mutual understanding of the goals and constraints of each party was not established upfront.

To be successful, Federal employees responsible for pursuing renewable energy projects must have a high-level understanding of the essential elements that make a viable deal for all parties to succeed. With this understanding, the Federal project lead can direct and manage specialists supporting a project and the development of a detailed project plan.

> A deal is a mutually beneficial business transaction among multiple parties.

The statement above is worth examining in further detail.

Mutually beneficial. Federal employees may not be well-versed in the motivations and constraints that drive private sector investment in project development. Project finance is structured with developers using project finance models, whereby each project is primarily financed by third-party debt and/or equity sources (not the developer's money) without the support of a corporate balance sheet. This means that the lenders and investors in a project rely solely on project revenues to generate returns.

Prior to bringing in investors, however, the developer directly bears the costs and risks of the upfront development of the project from concept to a fully documented and financeable deal. Because of the risk profile involved, and scarcity of development funds, developers are necessarily disciplined when expending development dollars and will ultimately choose to support projects they perceive to have the highest risk-adjusted return potential. Given this environment, Federal projects must present a competitive project opportunity to the development community to expect a strong, competitive response.

Business transaction. The business of large-scale renewable energy projects is relatively new, and therefore lacks commonly accepted definitions. This is an important barrier to the growth of renewable energy business in the United States under U.S. laws and business practices. The government's participation in this industry will help develop reliable, repeatable, business processes in the United States that investors can rely on. Successful business transactions require translation between the Federal sector and the private sector.

Multiple parties (commonly called counterparties). There are typically two primary parties that are required to execute a contract for renewable energy: the Federal Government and the developer. In practice, there are many relationships that have to be considered before a deal can close. Stakeholders and their roles and missions relative to the acquisition must be integrated into deal making. Failure to account for just one set of needs can cripple a deal. Multiple decision makers are essential to make a deal work and each have different, and potentially dynamic, points of view that must be continuously evaluated to reach the goal.

Examples of these counterparties that may be familiar to the reader include:

- Federal Government. Headquarters and regional elements, mission tenants, safety, environmental, BLM regarding land withdrawal.

- Developer. Sales team, division management, board level, shareholder advocates.

- Investors. Banks, middle-tier facilitators, insurance firms, private equity investors.

- Utility. Local utility, power marketers, public utility commissions, FERC, RTO/ISO.

- Lawyers. Representation for all parties.

- Local government. Elected community representatives, state and municipal officials, permitting agencies.

- Construction. Engineering, procurement, and construction contractors for the developer.

- Competitors. Fair federal procurement with a goal of no procurement protests.

- O&M Contractors. Responsible for project's O&M after project is constructed.

- Equipment suppliers. Provide the capital equipment for project.

- Feedstock suppliers. For projects requiring feedstock material (e.g., biomass).

Appendix E. 10-Step Project Development Framework Approach

It is possible to conceptualize a 10-step process to describe the major suggested steps in the project development framework approach for many methods of large-scale renewable energy project financing:

Step 1.

Establish project objectives using project fundamentals.

Step 2.

Conduct initial project assessment for fatal flaws using project development framework as a guide.

Step 3.

Make the decision to continue with an Incremental investment.

Step 4.

Perform further pre-development work deploying investment from Step 3, using project development framework.

Step 5.

Make the decision to continue with an incremental investment.

Step 6.

Conduct further pre-development work deploying investment from Step 5, using project development framework.

Step 7.

Make a Go/No Go decision to proceed with solicitation process initiation.

Step 8.

Pursue private partner through solicitation process, communicating project development framework.

Step 9.

Select and negotiate with awardee.

Step 10.

Conduct ongoing partnership with private sector developer as transaction counterparty.

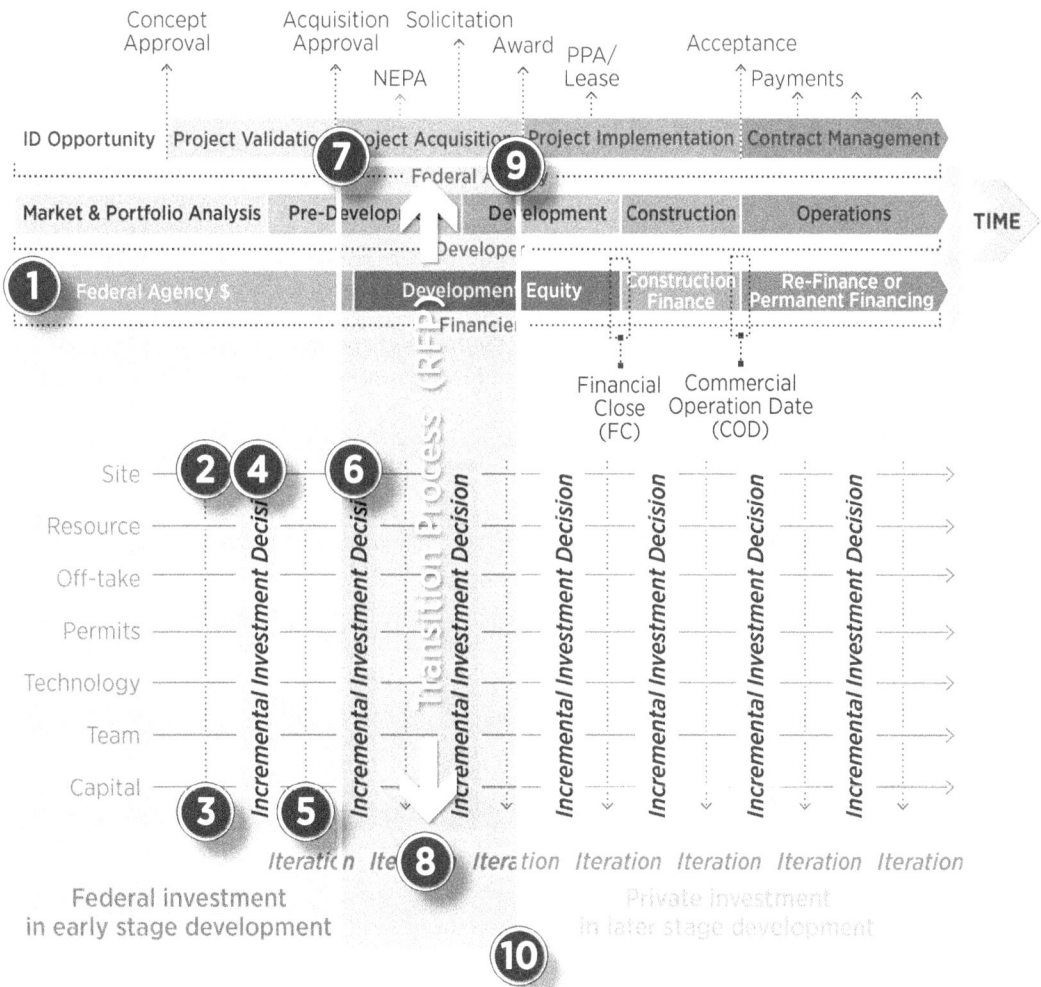

Figure E-1. Project development framework with 10-step process

Coordination of Government Phases and 10-Step Process

To help translate between the private and federal sector, this section describes the same steps above but delineates the steps in the respective government stages as outlined in Section III (Application of Project Development by a Federal Agency).

Stage 1. ID Opportunity

Step 1.

Establish project objectives using project fundamentals and Market and Portfolio Analysis.

Step 2.

Conduct initial project assessment for fatal flaws using pre-development steps of the development framework as a guide.

Step 3.

Make the decision to continue with an incremental investment including concept approval.

Stage 2. Project Validation

Step 4.

Perform further pre-development work deploying investment from Step 3, using pre-development steps of the project development framework.

Step 5.

Make the decision to continue with an incremental investment.

Step 6.

Conduct further pre-development work deploying investment from Step 5, using project development framework.

Stage 3. Project Acquisition

Step 7.

Make a Go/No Go decision on the competitive process initiation including acquisition plan approval.

Step 8.

Pursue private partner through competitive process, communicating project development framework.

Step 9.

Negotiate, select, and award.

Stage 4. Project Implementation

Step 10.

Conduct ongoing partnership with private sector developer as transaction counterparty.

These steps are more fully described below with general descriptions of each step. Note that when considering a large-scale renewable energy project using this outline of steps, experienced assistance and proven tools are essential to navigate the pitfalls of process and to inform key decisions throughout. Figure E-1 shows the steps labeled on the process diagram for ease of navigation.

Step 1.

Establish project objectives using project fundamentals and Market and Portfolio Analysis.

Project fundamentals are essential to establish a firm foundation with an objective purpose, which will generate the commitment of resources necessary to attract private investment—both project development capital and project finance capital.

Step 2.

Initial project assessment for fatal flaws using pre-development steps of project development framework.

Concentrating on the issues of Site, Resource, and Off-take, and the related issue of economics and bankability, conduct a high-level assessment of the seven project development framework subject areas with the purpose of identifying fatal flaws, significant areas of risk, and gaps of information. Engage experienced advisors to guide this process using established tools and professional opinion.

Step 3.

Incremental investment decision including concept approval.

With the assessment conducted using the project development framework, determine what has been learned, and using the output of a pro forma economic analysis determine if sufficient motivation exists to continue forward with an incremental investment. Confirm leadership buy-in and continued funding by obtaining concept approval. Set reasonable expectations. Direct investments to most efficiently take the project forward to the next decision point.

Step 4.

Further development work deploying investment from Step 3, using pre-development steps of project development framework as guide.

Invest time and resources to more fully develop the information within the project development framework, concentrating on the incremental investment suggested by the prior iteration. Use the project development framework to concentrate on verifying Site, Resource, and Off-take elements while continually seeking fatal flaws in Permits, Technology, Team, and Capital, and pursuing mitigating actions for risks across the project.

> **At this point, it should be clear that site conditions, resource, and off-take arrangements are economic and suggest a good investment.**

Step 5.

Incremental investment decision.

Does the project still meet the objectives? With the additional assessments conducted, determine what has been learned, and use the output of a pro forma economic analysis. Leadership should be briefed to determine whether sufficient motivation exists to continue forward with an incremental investment. At this point, it should be clear that Site, Resource, and Off-take arrangements are economic and suggest a good investment.

Further pre-development work deploying investment from Step 5, using project development framework as guide.

The result of this iteration should be a project plan that addresses and defines the known elements of the project development framework, in preparation for delivery to the private sector in a competitive acquisition process. References and documentation of the existing status should be in-hand, sufficient for developers to assess remaining risks and development activity as well as a clear and defined pathway to final approval and contract execution.

Go/No Go on competitive process document initiation including acquisition plan approval.

Based on assessment of the output of Step 6, a Go/No Go decision should be made. For the government, this decision falls under acquisition plan approval at the end of Stage 2. The project being offered should have risks that are acceptable and financeable by the private sector development community. For the government this would include issuing the competitive process document such as an RFP.

Pursue private partners through competitive process, communicating project development framework fundamentals.

Conduct a competitive acquisition of the project, acknowledging the continued development process and agreement to work as an active transaction partner to pursue project completion.

Negotiate, select, and award.

Qualify, select, negotiate, and award the opportunity.

Ongoing partnership with private sector developer as transaction counterparty.

Actively participate, with the support of ongoing Federal resources, the development of the project in conjunction with the selected private sector development team to carry the project forward into a completed deal, mutually beneficial to both parties, and/or other project participants or key stakeholders (such as the local utility). This step ends with commercial operation and acceptance.

Appendix F. Project Pro Forma Example

This appendix shows the output of a typical financial model for a 30-megawatt solar PV array. The model is provided simply to give the reader a sense of the depth for an investment-ready model and should be useful for government staff to gain a developer's perspective of a deal.

Typically, the government is focused on the price output from models such as the one here. Does the cost for the energy meet the government goals? The developer is typically looking at the return on investment (ROI) output. Does the project generate enough cash flow to pay off the project debt and yield a reasonable return on the equity invested into the project?

The Summary Information Sheet contains a snap shot of relevant project information that could be presented to obtain "board level" approval for an investment in the project. It demonstrates the sources of funds to implement the project, the income and expense over time, the likely rates of ROI, and the status of the project development data. Formats similar to this are commonly used for routine progress reports from pre-development to project financing. As the quality and accuracy of data improves, the range of likely returns will narrow until both risk and return are acceptable.

Pro formas are useful tools to perform "what if?" analyses by varying the key inputs. This informs the development team where to apply resources to improve the project.

Tools such as *pro formas* can be important for government portfolio development as projects can be readily compared. A consistent format with useful information in the same place on each report facilitates quick comparisons of project economics.

The data used to prepare the sample pro forma are meant to be realistic but not to represent current marketplace data. Actual inputs will vary depending on a project's market, technology, and geographic location.

The sample model inputs, shown in Table 1, are representative of many projects; they are the basis for finding whether a project is economic.

Solar PV cost input data are divided into two broad categories: capital and operation and maintenance (O&M) costs.

Capital costs are further categorized into direct and indirect costs. Direct costs are costs associated with the purchase of equipment: PV modules, inverter(s), balance of system (BOS), and installation costs. BOS costs are equipment costs that cannot be assigned to either the PV module or the inverter, and may include such costs as mounting racks, junction boxes, and wiring. Installation costs are the labor costs associated with installing the equipment. Indirect costs may include all other costs that are built into the price of a system, such as profit, overhead (including marketing), design, permitting, shipping, etc.

O&M costs are costs associated with a system after it is installed, and are categorized into fixed and variable O&M costs. Fixed O&M costs are costs that vary with the size of the system, and may include the cost of inverter replacements and periodic maintenance checks. Variable O&M costs vary with the output, use, and age of the system, and may be considered to be zero or very small for most PV systems.

The model uses the total installed cost, which is the sum of direct and indirect costs, to calculate the levelized cost of energy (LCOE). The LCOE is the present value of the energy price (e.g., cents/kilowatt hour) over the period of the sales contract. The LCOE of the renewable energy project is typically compared to the LCOE of continuing to purchase energy from the market over the same period to determine whether the project is cost effective.

EXPENSE INPUTS	INCOME INPUTS
How much will it cost to build?	What will the project produce?
Operation and maintenance?	How much will it produce?
How long will it take to build?	When will it occur?
How much do we need to borrow?	At what price?
How much will it cost to borrow the money?	What subsidies are available?
Fees for others (e.g. local utility)?	
Any taxes?	

Table 1. Financial Model Basic Elements[10]

[10] Expense categories and other inputs and outputs are representative of typical project costs, but are not meant to be exhaustive; projects should be reviewed and analyzed by qualified personnel.

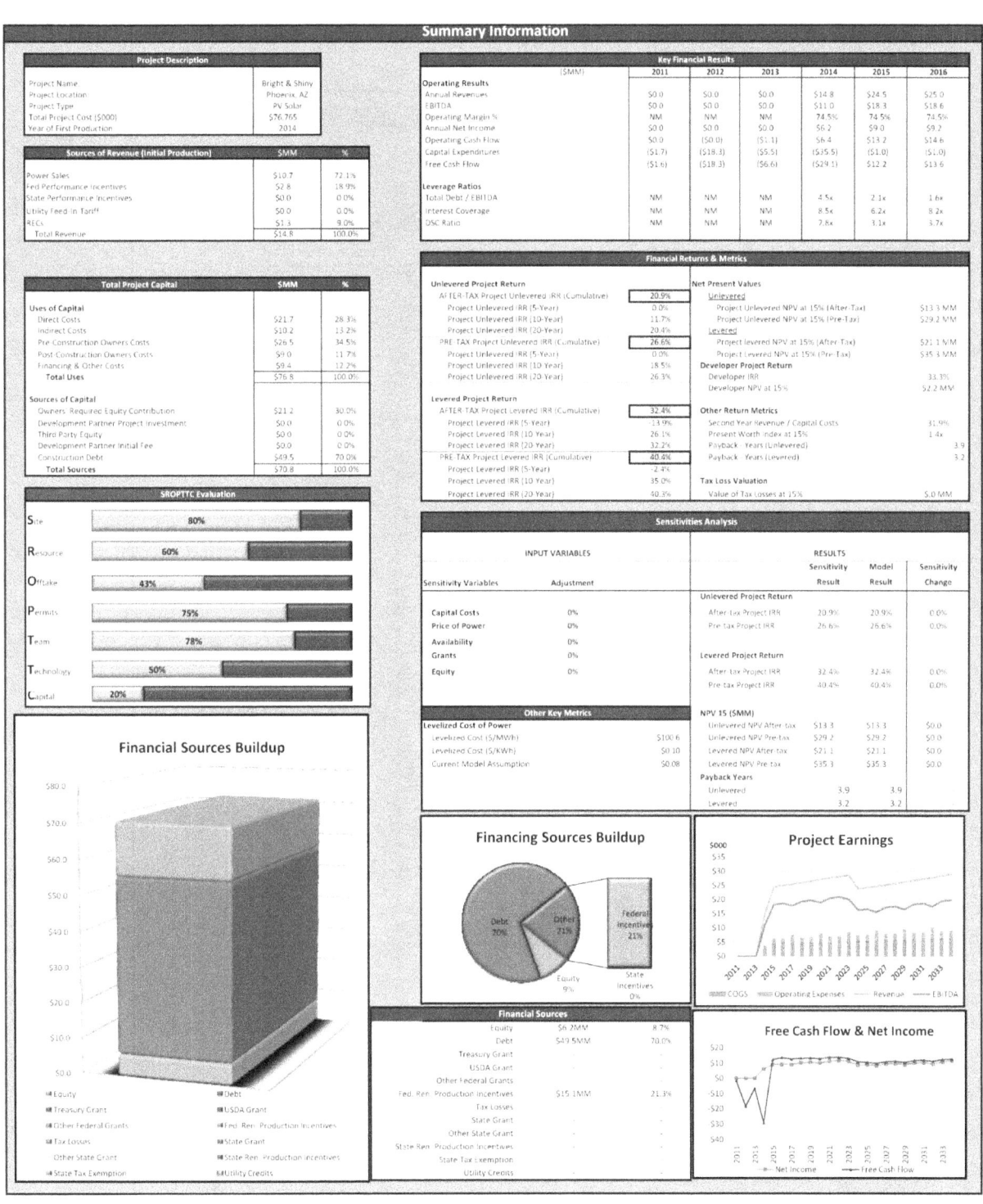

Summary Information Sheet

Solar PV Model (NREL)

Project Description		Financial		Construction Timing	
Project Name	Bright & Shiny	Total Capital Cost	$76,765	Construction Start	11/20/2012
Location	Phoenix, AZ	Equity Requirement	30.0%	Construction Period	18
Facility Assumptions		Debt Funding	70.0%	Plant Operations	5/20/2014
Gross Capacity Factor	86.0%	Interest Rate	L + 100		
Panel Availability	95.0%	Term	20 years		
Power Production		**Revenues**			
Generator Output (kW gross)	30,000	Wholesale Power Rate ($/kWh)	$0.08		
Parasitic Load (percentage)	5.0%	Renewable Energy Credits ($/kWh)	$0.01		
Generator Output (kW net)	28,500				

PV Solar Financial Model - Key Inputs

PROJECT AND OPERATING ASSUMPTIONS

Project Description ☑

Project Name	Bright & Shiny
Location	Phoenix, AZ
Project Type	PV Solar

Operating Assumptions ☑

Facility Assumptions

Gross Capacity Factor	86.0%
Annual Hours	8,760
Annual Available Hours	7,534
Panel Availability	95.0%
Annual Operating Hours	7,157
Net Power to Utility (kW)	28,500
Annual Net Power Output (kWh)	203,972,220

Power

Load Design (kW)	30,000
Parasitic Load Percentage	5.0%
Total Parasitic Load (kW)	1,500
Net Output to Utility (kW)	28,500

Construction Assumptions ☑

Pre-Construction Start Date	11/20/2011
Pre-Construction Period (months)	12
Construction Start Date	11/20/2012
Construction Period (months)	18
Plant Operations	5/20/2014
Construction Drawdown Schedule	Sculpted

Pre-Construction Owners' Costs Timing ☑

Category	Start	End
Pre-Development	11/20/2011	3/20/2012
Engineering & Design	3/20/2012	11/20/2012
Site Acquisition	11/20/2011	12/20/2011
Legal	11/20/2011	11/20/2012
Construction Working Capital Cash	11/20/2011	12/20/2011
Permitting	11/20/2011	11/20/2012

Debt Service Reserve ☑

Years of Plant Operations Before Reduction	2
Percentage of Mandatory Repayments	100.0%
Percentage of Annual Interest	100.0%

Capital Cost Estimate Assumptions ☑

Date of Capital Cost Estimates	1/1/2010
Annual Capital Cost Escalation Rate	3.0%

Other Pre-Construction Costs ($MM) ☑

Other Pre-Construction Costs ($MM)	$1.0
GIS Site Layout	$1.0

(Inputs) 1 of 4

Solar PV Model (NREL)

Interconnect Agreement	$1.0
Geotech Studies	$1.0
Foundation Study/Design	$1.0
Landscape and Visuals Permitting	$1.0
Total	$6.0

Panel Purchase Assumptions	
Lead Time (months)	5
Months Required Prior to End of Construction	10
Required Equipment Delivery Date	7/20/2013
Long Lead Equipment Cost ($MM)	$5.0
Down Payment %	10.0%
Down Payment Date	2/20/2013

INCOME STATEMENT ASSUMPTIONS

Revenue Assumptions (2011 Dollars)	
Revenue Pricing	
Wholesale Power Rate ($/kWh)	$0.08
Renewable Energy Credits ($/kWh)	$0.01
Revenue Escalation	2.0%
Interest Income %	2.0%
Carry Tax Losses?	Yes

Variable Cost Assumptions (2011 Dollars)	
Grid Connect Fee ($/kWh)	$0.01
Maintenance Fee ($/kWh)	$0.01
Other ($/kWh)	$0.00
Other ($/kWh)	$0.00

Tax & Depreciation Assumptions		
Federal Income Tax Rate		30.0%
State Income Tax		4.0%
Property Tax		
Total Property Value ($000)		$31,863
Property Exemption Rate		90.0%
Property Tax Basis ($000)		$3,186
Property Tax Rate		0.0%
Depreciation	*Book*	*Tax*
Owner's Costs	20 yr, S/L amort	10 yr MACRs
Interest During Construction	20 yr, S/L amort	20 yr MACRs
Financing Fees	20 yr, S/L amort	20 yr MACRs
Directs & Indirect Project Costs	20 yr, S/L amort	7 yr MACRs
Annual Maintenance Capex	20 yr, S/L amort	7 yr MACRs

Land and Maintenance	
Site Lease	Yes
Site Lease Cost	$500,000
Major Panel Maintenance Expense ($MM)	$1.00
Years between Major Maintenance	3

Other Operating Costs (annual)	
Direct Labor	$50,000
Materials	$20,000
Site Lease	500,000
Other Leases	80,000
Metering	150,000
Insurance	750,000
Regular Maintenance	100,000
Taxes and Fees	0
Routine O&M	50,000
Spare Parts	0
Other	0
Total	$1,700,000
Expense Escalation Rate	2.0%

Working Capital Assumptions	
Days Revenue in Product Sales A/R	30

(Inputs) 2 of 4

Solar PV Model (NREL)

Days Revenue in Prepaids A/R	0
Days Revenue in Other A/R	0
Turnover	365
Days COGS in A/P	28
Days Total Expenses in Accrued Payroll	11
Days Total Expenses in Accrued Liabilities	11
Operations Minimum Cash	$1.0

FINANCING ASSUMPTIONS

Financial
Project Costs (Escalated)
Total Project Capital Cost	$76,765
Hard Costs	$31,863
Owners' Costs	$35,500
Financing Costs	$9,402

Construction Funding
Equity Requirement	30.0%
Debt Funding	70.0%
Pricing on Construction Loan	L + 250
Amortization for Construction Loan	5.0%
Cash Sweep for Construction Loan	50.0%

Refinancing Debt Assumptions
Take-out Year	2017
Pricing on Refinancing Debt	L + 100
Amortization for Refinancing Loan	5.0%
Cash Sweep for Refinancing Debt	20.0%

Equity Distribution Percentage	100.0%
NPV Discount Rate	15.0%
Debt Placement Fee %	3.0%
Equity Placement Fee %	3.0%

Incentives Programs
NPV Rate for Analysis	15.0%

Federal
A. Grants
1. Treasury Grant	No
Cash Grant as % of Qualifying Equipment Costs	30.0%
Grant Max as % of Equity	100.0%
Cash Grant Amount ($MM)	$9.6
Grant Percentage to Equity at COD	100.0%
Grant Amount to Debt Principle at COD	0.0%
% Decrease of Depeciable Basis from Grant	50.0%
Timing	May-14
2. USDA Grant	No
Cash Grant as % of Qualifying Equipment Costs	25.0%
Grant Max as % of Equity	100.0%
Cash Grant Amount	$8.0
Grant Percentage to Equity at COD	100.0%
Timing	May-14
3. Other Federal Grants	No
Cash Grant Amount ($MM)	$5.00
Timing of Receipt (year)	2010

B. Performance Incentives
1. Renewable Energy Production Incentives	Yes
Maximum Generation Value ($/kWh)	$0.021/kWh
Maximum Percentage of Total Generation	100.0%
Term (years)	10
End Date (year)	2024

C. Tax Depreciation
1. Tax Depreciation on Equipment	7 yr MACRs

State
A. Grants
1. Development Grant	Yes
Cash Grant Amount ($MM)	$2.00
Timing of Receipt (year)	2010
2. Other Renewable Grants	No
Cash Grant Amount ($MM)	$2.00
Timing of Receipt (year)	2015

B. Performance Incentives

(Inputs) 3 of 4

Solar PV Model (NREL)

1. State Renewable Energy Production Incentives	No
Maximum Generation Value ($/kWh)	$0.011/kWh
Maximum Percentage of Total Generation	100.0%
Term (years)	6
End Date (year)	2020
C. Tax Exemptions	
1. State Tax Exemption	No
Percent of State Tax Exemption	100.0%
D. Utility Credits	
1. Feed In Tariff	No
Amount Above Wholesale Rates ($/kWh)	$0.030/kWh
Maximum Percentage of Total Generation	100.0%
Term (years)	8
End Date (year)	2022

Project Cost Assumptions (2010 Dollars)	☑
Direct Capital Costs	
Panels & Equipment	$4.5
Foundation	1.0
Interconnect	7.0
M&E Contractor Works	2.0
Grid Connection	1.0
Cable/Box	0.7
Other 1	1.0
Other 2	2.0
Contingency	0.0
Total Direct Costs	$19.2
Indirect Capital Costs	
Site Work	$2.0
Structural Steel	2.0
Insulation	2.0
Equipment Erection	2.0
Other	1.0
Contingency	0.0
Total Indirect Costs	$9.0
Pre-Construction Owners' Costs	
Pre-Development	$1.0
Panels & Equipment Down Payment	$0.5
Engineering & Design	10.0
Other Pre-Construction Costs	6.0
Site Acquisition	0.0
Legal	5.0
Construction Working Capital Cash	2.0
Permitting	2.0
Total Pre-Construction Owners' Costs	$26.5
Post-Construction Owners' Costs	
License Fees	$3.0
Development Fee	5.0
Startup Costs	1.0
Total Post-Construction Owners' Costs	$9.0
Financing Costs	
Debt Service Reserve	$5.4
Equity Placement Fees	0.6
Debt Placement Fees	1.5
Interest During Construction	1.9
Total Financing Costs	$9.4
Total Project Costs ($MM)	$73.1
Annual Maintenance Capex	$1.0

Solar PV Model (NREL)

Income Statement

			0.0%	0.0%	0.0%	61.6%	100.0%
Operational Factor			0.0%	0.0%	0.0%	61.6%	100.0%
Revenue Escalation Factor			1.00	1.02	1.04	1.06	1.08
Expense Escalation Factor			1.00	1.02	1.04	1.06	1.08
(Dollars in Millions)			2011	2012	2013	2014	2015
Revenue							
Power Sales	*rate:*	$0.080	$0.0	$0.0	$0.0	$10.7	$17.7
Fed Performance Incentives	*rate:*	$0.021	$0.0	$0.0	$0.0	$2.8	$4.6
State Performance Incentives	*rate:*	$0.000	$0.0	$0.0	$0.0	$0.0	$0.0
Utility Feed-In-Tariff	*rate:*	$0.000	$0.0	$0.0	$0.0	$0.0	$0.0
RECs			0.0	0.0	0.0	1.3	2.2
Total Revenue			$0.0	$0.0	$0.0	$14.8	$24.5
Cost of Goods Sold							
Grid Connect Fee			$0.0	$0.0	$0.0	$1.3	$2.2
Maintenance Fee			$0.0	$0.0	$0.0	$1.3	$2.2
Other			$0.0	$0.0	$0.0	$0.0	$0.0
Other			0.0	0.0	0.0	0.0	0.0
Total COGS			$0.0	$0.0	$0.0	$2.7	$4.4
Gross Profit			$0.0	$0.0	$0.0	$12.1	$20.1
Gross Margin			*NM*	*NM*	*NM*	*82.0%*	*82.0%*
Expenses							
Direct Labor			$0.0	$0.0	$0.0	$0.0	$0.1
Materials			0.0	0.0	0.0	0.0	0.0
Site Lease			0.0	0.0	0.0	0.3	0.5
Other Leases			0.0	0.0	0.0	0.1	0.1
Metering			0.0	0.0	0.0	0.1	0.2
Insurance			0.0	0.0	0.0	0.5	0.8
Regular Maintenance			0.0	0.0	0.0	0.1	0.1
Taxes and Fees			0.0	0.0	0.0	0.0	0.0
Routine O&M			0.0	0.0	0.0	0.0	0.1
Spare Parts			0.0	0.0	0.0	0.0	0.0
Other			0.0	0.0	0.0	0.0	0.0
Total Expenses			$0.0	$0.0	$0.0	$1.1	$1.8
EBITDA			$0.0	$0.0	$0.0	$11.0	$18.3
EBITDA Margin			*NM*	*NM*	*NM*	*74.5%*	*74.5%*
Depreciation and Amortization			0.0	0.0	0.0	(1.0)	(2.6)
Operating Income			$0.0	$0.0	$0.0	$10.1	$15.7
Interest Income			$0.0	$0.0	$0.0	$0.1	$0.1
Interest Expense			0.0	0.0	0.0	(1.3)	(2.9)
Pre-tax Income			$0.0	$0.0	$0.0	$8.9	$12.8
Income Tax			(0.0)	(0.0)	(0.0)	(2.7)	(3.9)
Net Income			$0.0	$0.0	$0.0	$6.2	$9.0

Balance Sheet

(Dollars in Millions)	2011	2012	2013	2014	2015
ASSETS					
Cash & Equivalents	$2.0	$2.1	$2.6	$9.7	$1.0
Debt Service Reserve	0.0	1.8	2.4	5.4	5.4
Deferred Tax Asset	0.0	0.0	0.0	(0.9)	(3.3)

(Financials) 1 of 3

Solar PV Model (NREL)

Accounts Receivable

Product Sales, Net	$0.0	$0.0	$0.0	$1.2	$2.0
Prepaids	0.0	0.0	0.0	0.0	0.0
Other	0.0	0.0	0.0	0.0	0.0

Inventory

Reagent/Chemicals/Other	$0.0	$0.0	$0.0	$0.0	$0.0
Other	0.0	0.0	0.0	0.0	0.0
Prepaid Expenses and Other	0.0	0.0	0.0	0.0	0.0
Total Current Assets	**$2.0**	**$3.8**	**$5.0**	**$15.4**	**$5.2**
Land, Cost	$0.0	$0.0	$0.0	$0.0	$0.0
Property & Equipment, Cost	0.0	1.2	6.5	31.9	31.9
Financing Fees & IDC, Cost	0.0	0.8	2.1	4.0	4.0
Other Development Spend, Cost	1.7	18.0	18.0	27.0	27.0
Maintenance Capital, Cost	0.0	0.0	0.0	0.0	1.0
Accumulated Depreciation & Amortization	0.0	0.0	0.0	(1.0)	(3.5)
Net Assets	**$1.7**	**$20.0**	**$26.7**	**$61.9**	**$60.4**

Other Assets

Deposits and Other Assets	$0.0	$0.0	$0.0	$0.0	$0.0
Total Other Assets	**$0.0**	**$0.0**	**$0.0**	**$0.0**	**$0.0**
Total Assets	**$3.7**	**$23.9**	**$31.6**	**$77.3**	**$65.5**

LIABILITIES & EQUITY

Revolving Credit Facility	$0.0	$0.0	$0.0	$0.0	$0.0
Accounts Payable	0.0	0.0	0.0	0.2	0.3
Accrued Payroll	0.0	0.0	0.0	0.0	0.0
Accrued Liabilities	0.0	0.0	0.0	0.0	0.1
Total Current Liabilities	**$0.0**	**$0.0**	**$0.0**	**$0.2**	**$0.4**
Debt	$0.0	$16.7	$22.1	$49.5	$37.8
Other Non-Current Liabilities	0.0	0.0	0.0	0.0	0.0
Total Long-Term Liabilities	**$0.0**	**$16.7**	**$22.1**	**$49.5**	**$37.8**
Total Liabilities	**$0.0**	**$16.7**	**$22.1**	**$49.8**	**$38.2**
Equity Project Financing	$3.7	$7.1	$9.5	$21.2	$21.2
Retained Earnings	0.0	0.0	0.1	0.1	6.3
Contributions / (Distributions)	0.0	0.0	0.0	0.0	(9.2)
Current Period Income (Loss)	0.0	0.0	0.0	6.2	9.0
Total Stockholders' Equity	**$3.7**	**$7.2**	**$9.6**	**$27.5**	**$27.3**
Total Liabilities & Equity	**$3.7**	**$23.9**	**$31.6**	**$77.3**	**$65.5**
Balance Check	*$0.0*	*$0.0*	*$0.0*	*$0.0*	*$0.0*

Balance Sheet Assumptions

	2011	2012	2013	2014	2015
Accounts Receivable Assumptions					
Product Sales, Net	$0.0	$0.0	$0.0	$1.2	$2.0
Days Revenue in Product Sales A/R	*30*	*30*	*30*	*30*	*30*
Prepaids	$0.0	$0.0	$0.0	$0.0	$0.0
Days Revenue in Prepaids A/R	*0*	*0*	*0*	*0*	*0*
Other	$0.0	$0.0	$0.0	$0.0	$0.0
Days Revenue in Other A/R	*0*	*0*	*0*	*0*	*0*

(Financials) 2 of 3

LARGE-SCALE RENEWABLE ENERGY GUIDE   59

Solar PV Model (NREL)

Inventory Assumptions					
Inventory	$0.0	$0.0	$0.0	$0.0	$0.0
Turnover	*365*	*365*	*365*	*365*	*365*
Accounts Payable	$0.0	$0.0	$0.0	$0.2	$0.3
Days COGS in A/P	*28*	*28*	*28*	*28*	*28*
Accrued Expense Assumptions					
Accrued Payroll	$0.0	$0.0	$0.0	$0.0	$0.0
Days Total Expenses in Accrued Payroll	*11*	*11*	*11*	*11*	*11*
Accrued Liabilities	$0.0	$0.0	$0.0	$0.0	$0.1
Days Total Expenses in Accrued Liabilities	*11*	*11*	*11*	*11*	*11*

Statement of Cash Flow

(Dollars in Millions)	2011	2012	2013	2014	2015
Cash Flow from Operations					
Net Income	$0.0	$0.0	$0.0	$6.2	$9.0
Add: Depreciation & Amortization	0.0	0.0	0.0	1.0	2.6
Less: Capitalized Interest Expense	0.0	(0.1)	(1.1)	(0.7)	0.0
Add: Increase in Deferred Tax Liabilities	(0.0)	(0.0)	(0.0)	0.9	2.3
Working Capital Change					
Accounts Receivable	$0.0	$0.0	$0.0	($1.2)	($0.8)
Inventory	0.0	0.0	0.0	(0.0)	(0.0)
Other Current Assets	0.0	0.0	0.0	0.0	0.0
Accounts Payable	0.0	0.0	0.0	0.2	0.1
Accrued Expenses	0.0	0.0	0.0	0.0	0.0
Total Cash Flow from Operations	$0.0	($0.0)	($1.1)	$6.4	$13.2
Cash Flow from Investing					
Land Acquisitions	$0.0	$0.0	$0.0	$0.0	$0.0
Property & Equipment Purchases	0.0	(1.2)	(5.3)	(25.3)	0.0
Other Development Costs	(1.7)	(17.0)	(0.2)	(10.2)	0.0
Maintenance Capital	0.0	0.0	0.0	0.0	(1.0)
Other Noncurrent Assets & Liabilities	0.0	0.0	0.0	0.0	0.0
Total Cash Flow from Investing	($1.7)	($18.3)	($5.5)	($35.5)	($1.0)
Free Cash Flow	($1.6)	($18.3)	($6.6)	($29.1)	$12.2
Cash Flow from Financing					
Line of Credit	$0.0	$0.0	$0.0	$0.0	$0.0
Equity Project Financing	3.7	3.5	2.3	11.8	0.0
Debt	0.0	16.7	5.4	27.5	(11.7)
Debt Service Reserve	0.0	(1.8)	(0.6)	(3.0)	0.0
Refinancing Fee	0.0	0.0	0.0	0.0	0.0
Grant Proceeds	0.0	0.0	0.0	0.0	0.0
Contributions	0.0	0.0	0.0	0.0	0.0
Distribution	0.0	0.0	0.0	0.0	(9.2)
Total Cash Flow from Financing	$3.7	$18.4	$7.2	$36.2	($21.0)
Increase / (Decrease) in Cash	$2.0	$0.0	$0.5	$7.1	($8.7)
Beginning Cash	0.0	2.0	2.1	2.6	9.7
Ending Cash	$2.0	$2.1	$2.6	$9.7	$1.0

(Financials) 3 of 3

Solar PV Model (NREL)

Operating Summary

(Dollars in Millions)				2011	2012	2013
Power Sales						
Volume (MW)				28.5	28.5	28.5
Availability				0.0%	0.0%	0.0%
Annual Hours				8,760	8,760	8,760
Total Output				0	0	0
Price ($/MWh)	rate:		$0.08	$80.00	$81.60	$83.23
Power Sales				$0.0	$0.0	$0.0
Other Sources of Revenue						
Federal Performance Incentives				$0.0	$0.0	$0.0
State Performance Incentives				$0.0	$0.0	$0.0
Utility Feed-In-Tariff				$0.0	$0.0	$0.0
RECs				0.0	0.0	0.0
Total Other Revenue				$0.0	$0.0	$0.0
Total Revenue				$0.0	$0.0	$0.0
Cost of Goods Sold						
Grid Connect Fee				$0.0	$0.0	$0.0
Maintenance Fee				0.0	0.0	0.0
Other				0.0	0.0	0.0
Other				0.0	0.0	0.0
Total COGS				$0.0	$0.0	$0.0
Gross Profit				$0.0	$0.0	$0.0
Gross Margin				NM	NM	NM
Operating Expenses				$0.0	$0.0	$0.0
EBITDA				$0.0	$0.0	$0.0
EBITDA Margin				NM	NM	NM
Adjustments to EBITDA to get to Operating Cash Flow						
Changes in Net Working Capital				$0.0	$0.0	$0.0
Net Interest				0.0	(0.0)	(1.1)
Cash Taxes				(0.0)	(0.0)	(0.0)
Total Adjustments				$0.0	($0.0)	($1.1)
Operating Cash Flow				$0.0	($0.0)	($1.1)
Capital Expenditures				($1.7)	($18.3)	($5.5)
Free Cash Flow				($1.6)	($18.3)	($6.6)

(Financial Analysis) 1 of 4

Solar PV Model (NREL)

Project Return Metrics

(Dollars in Millions)	2011	2012	2013
Unlevered Cash Flow			
Free Cash Flow	($1.6)	($18.3)	($6.6)
Add: Distributed Tax Grant Proceeds	0.0	0.0	0.0
Add: Interest Expense and Capitalized Interest	(0.0)	0.0	1.1
Less: Interest Tax Shield	0.0	0.0	0.0
Total After-Tax Unlevered Cash Flow	**($1.7)**	**($18.3)**	**($5.5)**
Negative Cash Flows	*($1.7)*	*($18.3)*	*($5.5)*
Postitive Cash Flows	*0.0*	*0.0*	*0.0*
Cummulative Cash Flow	*0.0*	*0.0*	*0.0*
Payback	*0.0*	*0.0*	*0.0*
Annual IRR	*NM*	*NM*	*NM*
Add: Cash Taxes	0.0	0.0	0.0
Total Pre-tax Unlevered Cash Flow	**($1.7)**	**($18.3)**	**($5.5)**

Project Unlevered NPV at 15% (After-Tax)	$13.3
Project Unevered IRR (After-Tax)	20.9%
Project Unlevered NPV at 15% (Pre-Tax)	$29.2
Project Unevered IRR (Pre-Tax)	26.7%
Capital Costs	$76.8
Payback - Years	*3.9*

(Dollars in Millions)	2011	2012	2013
Levered Cash Flows to Equity			
Equity Constributions During Construction	($3.7)	($3.5)	($2.3)
Less: Contributions	0.0	0.0	0.0
Add: Distributed Tax Grant Proceeds	0.0	0.0	0.0
Add: Equity Distributions from Cash Flow	0.0	0.0	0.0
Total After-Tax Levered Cash Flows to Equity	**($3.7)**	**($3.5)**	**($2.3)**
Postitive Cash Flows	*$0.0*	*$0.0*	*$0.0*
Cummulative Cash Flow	*0.0*	*0.0*	*0.0*
Payback	*0.0*	*0.0*	*0.0*
Annual IRR	*NM*	*NM*	*NM*
Add: Cash Taxes	0.0	0.0	0.0
Total Pre-tax Levered Cash Flow	**($3.7)**	**($3.5)**	**($2.3)**

Project levered NPV at 15% (After-Tax)	$21.1
Project Levered IRR (After-Tax)	32.4%
Project levered NPV at 15% (Pre-Tax)	$35.3
Project Levered IRR (Pre-Tax)	40.4%
Equity Required	$21.2
Payback - Years	*3.2*

(Financial Analysis) 2 of 4

Solar PV Model (NREL)

Additional Project Return Items

(Dollars in Millions)		Nov-11	Dec-11	Jan-12
Cash Flows to Developer				
Developer Equity Contributions		($2.8)	($0.8)	($0.8)
Developer Repayment at Financial Close		0.0	0.0	0.0
Development Fee		0.0	0.0	0.0
Total Cash Flows to Developer		**($2.8)**	**($0.8)**	**($0.8)**
Developer IRR	33.3%			
Developer NPV at 15%	$2.2			

Other Return Metrics				
Second Year Revenue / Capital Costs	31.9%			
Present Worth Index at 15%	1.4x			

Tax Loss Valuation		2011	2012	2013
Potential Tax Losses		$0.0	$0.0	$0.0
Assumed Tax Rate		34.0%	34.0%	34.0%
Potential Value		$0.0	$0.0	$0.0
Discount		0.0%	0.0%	0.0%
Discounted Value		$0.0	$0.0	$0.0

Sale Value Exists if Tax Losses Not Carried Forward
Value of Tax Losses at 15% $0.0

(Financial Analysis) 3 of 4

Solar PV Model (NREL)

Credit Statistics

(Dollars in Millions)		2011	2012	2013
Financial Information				
Revolver		$0.0	$0.0	$0.0
Construction Loan		0.0	16.7	22.1
Refinancing Debt		0.0	0.0	0.0
Total Debt		0.0	16.7	22.1
Total Interest		0.0	0.0	0.0
EBITDA		0.0	0.0	0.0
Leverage Ratios				
Total Debt / EBITDA		NM	NM	NM
Interest Coverage		NM	NM	NM
DSC Ratio		NM	NM	NM

Leverage Ratios

Avg. Total Debt/EBITDA	1.4x
Min Interest Coverage	6.2x
Avg. DSC Ratio	7.6x

(Financial Analysis) 4 of 4

Solar PV Model (NREL)

Sources and Uses

Sources	Total	%
Equity		
Owners' Required Equity Contribution	$21.2	30.0%
Development Partner Project Investment	0.0	0.0
Third Party Equity	0.0	0.0
Development Partner Initial Fee	0.0	0.0
Total	**$21.2**	**30.0%**
Debt		
Construction Debt	$49.5	70.0%
Other Debt	0.0	0.0
Total Debt	**$49.5**	**70.0%**
Total Sources	**$70.8**	**100.0%**

Uses ($MM)	Total	%
Direct Capital Costs		
Panels & Equipment	$5.1	6.6%
Foundation	1.1	1.5
Interconnect	7.9	10.3
M&E Contractor Works	2.3	2.9
Grid Connection	1.1	1.5
Cable/Box	0.8	1.0
Other 1	1.1	1.5
Other 2	2.3	2.9
Contingency	0.0	0.0
Total Direct Costs	**$21.7**	**28.3%**
Indirect Capital Costs		
Site Work	$2.3	2.9%
Structural Steel	$2.3	2.9
Insulation	$2.3	2.9
Equipment Erection	$2.3	2.9
Other	$1.1	1.5
Contingency	$0.0	0.0
Total Indirect Costs	**$10.2**	**13.2%**
Pre-Construction Owners' Costs		
Pre Development	$1.0	1.3%
Engineering & Design	10.0	13.0
Site Acquisition	0.0	0.0
Legal	5.0	6.5
Permitting	2.0	2.6
Total	**$26.5**	**34.5%**
Post-Construction Owners' Costs		
License Fees	$3.0	3.9%
Development Fee	5.0	6.5
Construction Working Capital Cash	2.0	2.6
Startup Costs	1.0	1.3
Total	**$9.0**	**11.7%**
Financing & Other Costs		
Debt Service Reserve	$5.4	7.0%
Equity Placement Fees	0.6	0.8
Debt Placement Fees	1.5	1.9
Interest During Construction	1.9	2.5
Total Financing Costs	**$9.4**	**12.2%**
Total Uses	**$76.8**	**100.0%**

(Source & Uses) 1 of 1

Solar PV Model (NREL)

Monthly Construction Draw Schedule

				2011	2011	2012	2012	2012
Year				2011	2011	2012	2012	2012
Date				Nov-11	Dec-11	Jan-12	Feb-12	Mar-12
Project Month				1	2	3	4	5
Construction Month				--	--	--	--	--
Libor				3.000%	3.000%	3.150%	3.150%	3.150%

		Project Month							
(Dollars in Millions)		Begin	End		Nov-11	Dec-11	Jan-12	Feb-12	Mar-12
Pre-Construction Owners' Costs									
Pre-Development	$1.0			$1.0	$0.3	$0.3	$0.3	$0.3	$0.0
Turbine Down Payment	$0.5			$0.5	$0.0	$0.0	$0.0	$0.0	$0.0
Engineering & Design	10.0			$10.0	0.0	0.0	0.0	0.0	1.3
Other Pre-Development Costs				$0.0					
Site Acquisition	0.0			$0.0	0.0	0.0	0.0	0.0	0.0
Legal	5.0			$5.0	0.4	0.4	0.4	0.4	0.4
Construction Working Capital Cash	2.0			$2.0	2.0	0.0	0.0	0.0	0.0
Permitting	2.0			$2.0	0.2	0.2	0.2	0.2	0.2
Total	**$26.5**				**$2.8**	**$0.8**	**$0.8**	**$0.8**	**$1.8**
Post-Construction Owners' Costs									
License Fees	$3.0	30	30		$0.0	$0.0	$0.0	$0.0	$0.0
Development Fee	5.0	30	30		0.0	0.0	0.0	0.0	0.0
Startup Costs	1.0	30	30		0.0	0.0	0.0	0.0	0.0
Total	**$9.0**				**$0.0**	**$0.0**	**$0.0**	**$0.0**	**$0.0**
Financing Costs									
Debt Service Reserve	$5.4				$0.0	$0.0	$0.0	$0.0	$0.0
Equity Placement Fees	0.6	3.0%			0.0	0.0	0.0	0.0	0.0
Debt Placement Fees	1.5	3.0%			0.0	0.0	0.0	0.0	0.0
Interest During Construction	1.9	L + 250			0.0	0.0	0.0	0.0	0.0
Total Financing Costs	**$9.4**				**$0.0**	**$0.0**	**$0.0**	**$0.0**	**$0.0**
Direct Capital Costs									
Panels & Equipment	$4.5				$0.0	$0.0	$0.0	$0.0	$0.0
Foundation	1.0				0.0	0.0	0.0	0.0	0.0
Interconnect	7.0				0.0	0.0	0.0	0.0	0.0
M&E Contractor Works	2.0				0.0	0.0	0.0	0.0	0.0
Grid Connection	1.0				0.0	0.0	0.0	0.0	0.0
Cable/Box	0.7				0.0	0.0	0.0	0.0	0.0
Other 1	1.0				0.0	0.0	0.0	0.0	0.0
Other 2	2.0				0.0	0.0	0.0	0.0	0.0
Contingency	0.0				0.0	0.0	0.0	0.0	0.0
Total Direct Costs	**$19.2**				**$0.0**	**$0.0**	**$0.0**	**$0.0**	**$0.0**
Indirect Capital Costs									
Site Work	$2.0				$0.0	$0.0	$0.0	$0.0	$0.0
Structural Steel	2.0				0.0	0.0	0.0	0.0	0.0
Insulation	2.0				0.0	0.0	0.0	0.0	0.0
Equipment Erection	2.0				0.0	0.0	0.0	0.0	0.0
Other	1.0				0.0	0.0	0.0	0.0	0.0
Contingency	0.0				0.0	0.0	0.0	0.0	0.0
Total Indirect Costs	**$9.0**				**$0.0**	**$0.0**	**$0.0**	**$0.0**	**$0.0**
Total Construction Costs w/o Escalation	**$28.2**				**$0.0**	**$0.0**	**$0.0**	**$0.0**	**$0.0**
Escalation Rate	3.7				1.06	1.06	1.06	1.07	1.07
Total Construction Costs w/ Escalation	**$31.9**				**$0.0**	**$0.0**	**$0.0**	**$0.0**	**$0.0**
Total Funding Needs	**$70.8**				**$2.8**	**$0.8**	**$0.8**	**$0.8**	**$1.8**
Developer Equity Funding	$20.0	28.3%			$2.8	$0.8	$0.8	$0.8	$1.8
Developer Repayment at Financial Close	(20.0)	(28.3%)			0.0	0.0	0.0	0.0	0.0
Project Debt Funding	49.5	70.0%			0.0	0.0	0.0	0.0	0.0
Project Equity Funding	21.2	30.0			0.0	0.0	0.0	0.0	0.0
Total	**$70.8**	**100.0%**			**$2.8**	**$0.8**	**$0.8**	**$0.8**	**$1.8**
Debt Balance	$49.5				$0.0	$0.0	$0.0	$0.0	$0.0
Equity Balance	$21.2				$2.8	$3.7	$4.5	$5.3	$7.2
Grant Proceeds	$0.0				$0.0	$0.0	$0.0	$0.0	$0.0

Construction Schedule Options

Appendix G. Project Risk Assessment Template

The U.S. Army Energy Initiatives Task Force (EITF) serves as the Army's central management office to implement cost-effective large-scale renewable energy projects on Army installations leveraging industry financing. The EITF is an example of how one Federal agency has institutionalized and developed a repeatable process for large-scale renewable energy project development. FEMP, NREL, and the Army EITF have worked collaboratively in parallel on these processes.

This example is the Army's current Draft approach to methodically reviewing key project elements in an approach similar to the Project Development Framework described in Section II of this Guide.

Project Risk Assessment Template

Project Risk Assessment	
Mission/ Security	• How does project enhance energy security on Installation? • What are the possible impacts to Installation operations?
Economics	• What is the estimate of the baseline capital cost? • Have all other development costs been included? • What is the value of any REC's? • Is resource validation required? What is the status? • What is existing utility rate and alternative tariffs? • Does Economic Case Analysis (ECA) show cost savings for Army considering current and project utility rates?
Real Estate	• What is the approach and what authority is being used? • What are issues to obtaining required BLM agreement?
Regulatory (Legal)	• What are the regulatory limits for interconnection, net-metering? • What is the status of getting required PUC approvals?
Off-Take	• How much does installation use now and is this sufficient to consume all electricity? • If power is to be sold off the installation, have off-takers been identified? • What is the status of state RPS to drive demand?
Integration (Technical)	• What are the technical issues to connect to grid (e.g., substation, line capacity, etc.)? • What is the status of required interconnect or flow studies?
NEPA	• What are the major NEPA issues? • Who will implement NEPA and what is the timeline?
Acquisition	• What is acquisition strategy and timeline to implement? • What performance risks are there with the developer or other partners?

Project Risk Factors are reviewed on a regular basis to identify roadblocks and key issues for successful project development

Assistant Secretary of the Army (Installations, Energy & Environment)

Energy Initiatives Task Force   *Unclassified*

Figure G-1. Example Army project risk assessment framework. Source: U.S. Army

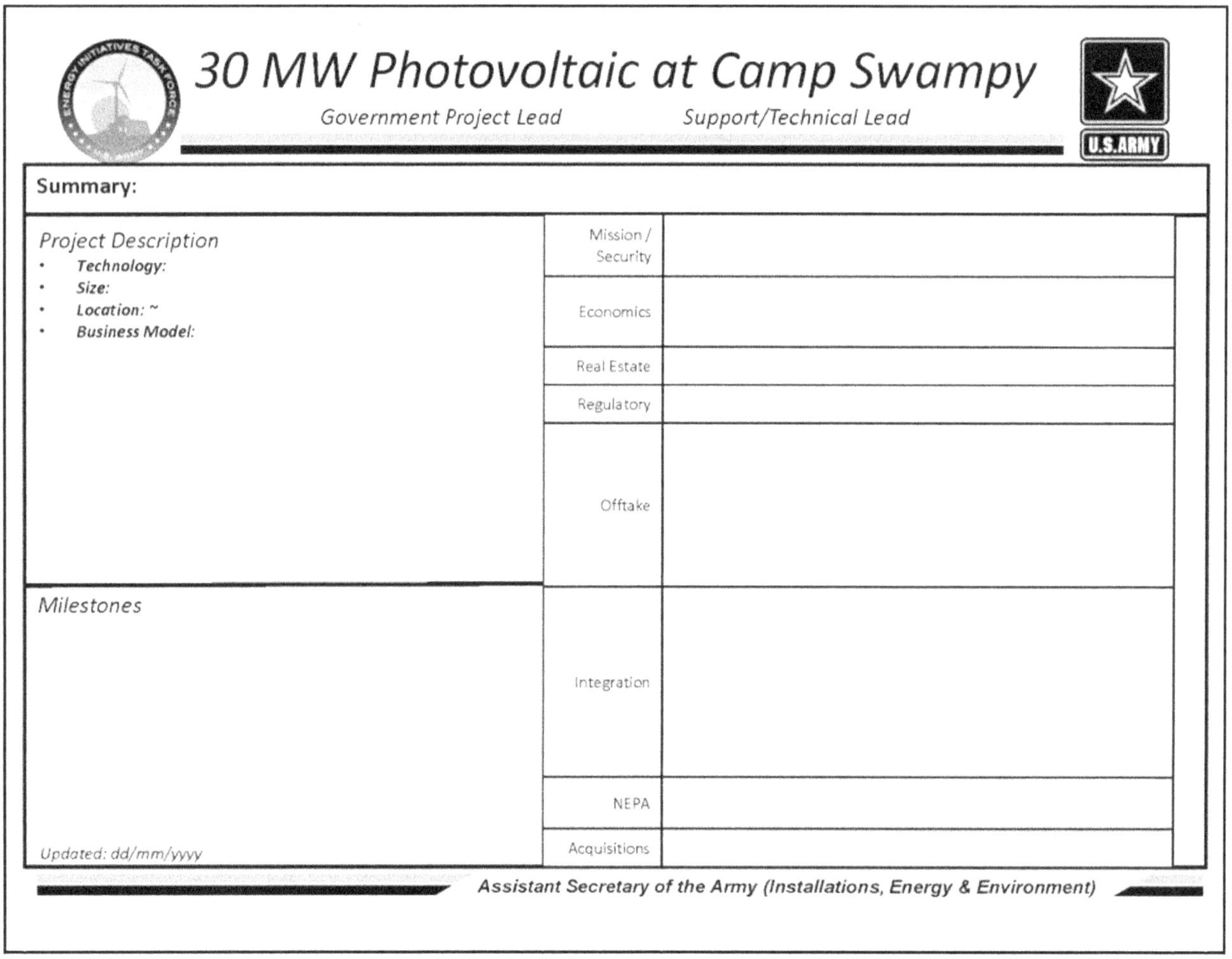

30 MW Photovoltaic at Camp Swampy

Government Project Lead　　　　*Support/Technical Lead*

Summary:

Project Description
- *Technology:*
- *Size:*
- *Location: ~*
- *Business Model:*

Milestones

Updated: dd/mm/yyyy

Mission / Security	
Economics	
Real Estate	
Regulatory	
Offtake	
Integration	
NEPA	
Acquisitions	

Assistant Secretary of the Army (Installations, Energy & Environment)

Figure G-2. Example Army project risk assessment template. Source: U.S. Army

Appendix H. Project Validation Report (DRAFT)

To provide Federal agencies with an example of a methodical approach to project validation (Stage 2 of a typical government process), this appendix includes an example of the project validation report template the U.S. Army Energy Initiatives Task Force (EITF) currently uses to document the analysis supporting the recommended course of action to the Army leadership.

Project Validation Report

The Project Validation Report contains all the project analysis and supporting data and documentation that has been completed on a renewable energy project. This document is used as the basis for the Agency to move forward with a solicitation. This document is updated on a regular basis and serves as the official record of the project development effort.

1. Executive Summary
   The Executive Summary is an overview of the Project Definition: its goals and objectives; the cost implications (appropriated and non-appropriated), the acquisition model, and the overall impact to the organization.
2. Recommended Project Description
   a. Resource and Technology Assessment
      i. System Overview and Assumptions
      ii. Integration with existing site infrastructure
   b. Energy Goal Impacts
      i. Impact on site energy security/goals
      ii. Impact on organization energy goals/mandates
   c. Leadership endorsement
   d. Summary of Alternatives Evaluated (detailed matrix to be included in an appendix)
3. Project Considerations
   a. Economics - Project Economic Analysis
      i. Conceptual Cost Estimation Overview
      ii. Financial Pro-forma
         1. Assumptions of pro-forma
         2. 3rd Party Investment Requirement
            a. NPV, IRR, Profitability
         3. Calculated Cost to the Organization
            a. LCOE and annual cost projections
            b. NPV, IRR
         4. Other Organization Costs
         5. Reliability premium and justification
         6. Risks and Sensitivity Analysis

Project Validation Report

    b. Real Estate - Real Property Land Agreements / Report of Availability
        i. Description of property, including any improvements
        ii. Real estate vehicle and terms & conditions
        iii. How will the site and proposed project incorporate into the approved Master Plan, if any?
        iv. United States Property Interest
        v. Attach maps

    c. Regulatory
        i. Description of State and Local Regulations
        ii. Description of Available Project Incentives
        iii. Requirements for Developer

    d. Off-take
        i. Market Area Analysis
            1. Utility(ies) Identification and Assessment
            2. Renewable Energy Market Analysis
            3. Transmission Capacities and availabilities
        ii. Likelihood of developer interest in project

    e. Technical Integration
        i. External connection issues and studies such as an Interconnection Assessment, including System Impact Study if available
        ii. Internal connection issues such as Expected impacts of integration into existing site infrastructure

    f. NEPA
        i. Environmental Assessment or Environmental Impact Study Results as defined by the National Environmental Policy Act

    g. Mission
        i. Site Demand and Reliability Requirements
        ii. Site Security: impact on site's existing buildings and utilities
        iii. Physical Limitations: description of any physical limitations that are present at the site, anything that could cause a problem to the development of the project
        iv. Transportation and Site Access: description of all access available to the installation. State any limitations that may exist and what will be done to facilitate any necessary changes, i.e. new roads or paving

    h. Acquisition
        i. Developer Requirements
        ii. Proposed Acquisition Strategy
        iii. Proposed Construction Management Plan
        iv. Contract Lifecycle Management Plan

4. Outreach Plan
    a. Key Stakeholders
    b. Outreach strategy
    c. Major Milestones
        i. Press Releases
        ii. Congressional Notifications
        iii. Industry Day Plans

Appendix I. Summary of Responses to Comments.

On June 1, 2012, DOE published a draft of the document titled "Developing Large-Scale Renewable Energy Projects at Federal Facilities using Private Capital" (Large RE Guide) and requested comments to solicit information and data from industry and stakeholders. The Federal Register notice identified seven specific issue areas on which DOE sought additional information. DOE received a total of thirteen timely submitted comments. Specifically, DOE received comments from six industry representatives and one industry trade association. The remaining comments were provided by representatives of two universities, one non-profit organization, one state government agency, one media group, and one member of the public. The commenters are identified at the end of this notice and all of the comments are available on the FEMP website http://www1.eere.energy.gov/femp/. DOE has considered all information received in the comments in developing the final version of the Large RE Guide.

The following section provides a summary of comments DOE received in response to the Federal Register notice. Generally, the comments can be grouped into six main issue areas: definitions, economics, risk, roles and responsibilities, schedule and timing, and editorial. Some comments were beyond the scope of the present version of the Large RE Guide and may be considered in future updates.

DOE received various comments regarding definitions of the types of renewable energy addressed in the Large RE Guide. Specifically, DOE received comments that the Large RE Guide did not address biodiesel refineries or general hydropower. DOE responded to these comments by clarifying that the types of renewable resources addressed in the Large RE Guide are those defined by law and Executive Order, as presented in the Introduction section of the Large RE Guide. Consistent with the definition of "renewable energy" in the Energy Policy Act of 2005, the Large RE Guide does not discuss hydroelectric generation capacity that is not achieved from either increased efficiency or additions of new capacity at an existing hydroelectric project. Also, the Introduction now clarifies that this document is focused on renewable energy for facilities, not biodiesel refineries for producing fuel. DOE has also referred to the definition of "renewable energy" as used in the Large RE Guide instead of referencing only electric energy generated from solar and wind. DOE received a comment that various different terms were used to describe the process detailed in the Large RE Guide. DOE addressed these comments by using the single term "Framework" to describe the process more consistently. DOE received a comment pointing out that the term "energy security" was not always understood by facility managers. A clarifying statement was added to the Baseline section of Section II, addressing energy security at a high level. It is beyond the scope of this document to discuss energy security in detail.

Regarding economics, one commenter recommended the creation of standards for payback period analyses. DOE did not

attempt to do so in the final version of the Large RE Guide in part because the key economic metric for large-scale renewable energy projects is levelized cost of electricity (LCOE) rather than payback. LCOE is addressed in Appendix F entitled Project Pro Forma Example. DOE also received two comments regarding the financing of power purchase agreements (PPAs). The first comment requested that each PPA be reviewed to determine whether subsidies were embedded in the PPA rate. This request is outside the scope of the Large RE Guide. However, DOE added further discussion to the off-take category of the Project Development section confirming that PPAs must meet government requirements, state laws, and procurement policies. The second comment asked DOE to mention more benefits of renewable energy projects. DOE reviewed the benefits that were already mentioned in the Large RE Guide and added long-term fixed price contract under Motivations in Figure 3 on the language barrier. A long term fixed price contract for energy provides certainty for planning and budget purposes.

DOE received many comments about how important it is for private developers to assess project risk. The final version of the Large RE Guide has added fuller discussions of Federal large renewable project risk and how that risk is balanced between the Federal Government and the private developer. A discussion of project risk has also been added to the Executive Summary, and is included in several highlight boxes. DOE received a comment requesting a discussion of ways the Federal Government can reward commercial developers who put time and effort into the early phases of these large Renewable projects without remuneration. DOE addressed this comment by explaining that an important Federal contracting principle is that firms that help develop requirements for a specific project generally are not eligible to bid on the acquisition resulting from those requirements absent a written waiver of a potential conflict of interest. However, firms usually may provide general information on technology, market conditions, and other relevant information without a conflict arising. This approach may be different than private project practice. In addition, DOE clarified that a developer that is eligible to bid on an acquisition can include in the price of the proposal the value of any initial work the developer puts into a project prior to winning the award.

The need for expertise in the early stages (which is the value that the companies provide) was addressed by recommending that the Federal Government use subject matter experts within the Government or hired to provide advice. These subject matter experts are not eligible to bid on the resulting project, absent a waiver. Suggestions requesting further detail on risks related to government contract terms, sunk development costs, and transactional costs will be considered for later versions of the Large RE Guide. A few comments that Federal agencies should provide existing data about potential projects to developers were addressed by encouraging disclosure of as much data as possible, as well as lessons learned from failed projects. Comments about the risks related to the uncertainties of compliance with the National Environmental Policy Act (NEPA) and of various permitting processes were addressed by adding further discussion of approaches to satisfy these requirements.

Regarding the area of roles and responsibilities, one comment stated that project success from a developer's point of view depends on having a market, a way to get to the market, and the ability to obtain all of the relevant permits. DOE agreed that the comment summarized key project elements and added the commenter's statement in Section II. DOE also received comments that renewable resource and other data collected by the Federal Government for the project meet industry standards. DOE addressed this comment by adding language in Section II and Appendix B to encourage the Federal Government to collect project data that meets industry standards in an effort to avoid duplication of effort and reduce risks. Several commenters sought clarification on the NEPA process and how NEPA requirements for projects involving environmental or cultural protection could impact costs and project feasibility. This version of the Large RE Guide clarifies the NEPA process, discusses the impact of the NEPA process on project feasibility, and includes discussion of the roles of the Federal Government and the project developer in this process. A comment stated that utilities may have unfair advantage for some projects. Specifically, the commenter was concerned that if the Federal Government looked at contracting directly with a utility under the sole-source authority covering local utilities, there would be less opportunity for developers except as subcontractors to the utility. The final document discusses in Appendix B3 Off-take the context for considering contracts with utilities.

In the area of schedule and timing, several commenters noted that project schedule was very important, and they sought more detail on the topic of schedule in general and in the discussion supporting the graphic in Figure 1. These comments and DOE's responses are described in detail below. As discussed in Section I, DOE explained that Figure 1 is a simplified concept illustration presented to capture steps and key milestones and not intended to provide detailed information on specific project schedules. DOE also noted that timelines may vary considerably due to the uniqueness of the project at issue. Several commenters indicated that the lack of a clear discussion of project timing for Federal large renewable projects was one area of difficulty and risk for industry. One comment said that adding a time horizon would make large Federal projects more attractive to the private sector. Another comment noted that key project elements are time-sensitive and may depend on incentives that expire. DOE acknowledges these comments and will work to improve this area in the next version of the Large RE Guide, as the subject is complex and beyond the resources available for this version. As large renewable projects are relatively new for many Federal agencies, exact timing and patterns are difficult to define at this time. DOE expects that with more experience, the schedules for these projects will become better defined.

There were comments that were not addressed in the final document for a variety of reasons. Some comments were compliments or statements describing the Large RE Guide that made no recommendation and thus required no response. Some comments in response to questions in the Federal Register notice affirmed that the Large RE Guide provided reasonable useful information, and thus required no response. One commenter included several comments that the Large RE Guide was too detailed and confusing, but these comments were balanced by more comments that supported the Large RE Guide and its level of detail. Several comments requested more detail on case studies, templates, and methodologies that could not be provided in this initial high level version of the document. These comments will be considered when updates of this document are made in the future. Many comments provided additional information on energy security that was too detailed for this high level document. However, these comments will be considered for inclusion in future updates of the document.

Finally, DOE received several comments highlighting key differences in early development stages from the industry's perspective as well as from the Federal model. DOE has addressed these comments by adding a more detailed discussion of key planning issues under Stage 1 in Section I and under Stage 1 in Section II.

List of Commenters –

1 – Fred Morse, Abengoa (Industry)

2 – Tom Clements, Alliance for Nuclear Accountability (Non-Profit Organization)

3 – Bill Elliott, Contractor at Ft. Belvoir (Government Agency, Contractor User)

4 – Donna K. Albert, Energy Project Manager, Energy Program, WA, (State Government)

5 – David Bransby, Professor of Energy Crops and Bioenergy, Auburn University, (Education)

6 – Michael Theroux, terutalk.com (Media)

7 – John Lehto (Private Citizen)

8 – Edward Lovelace, Free Flow Power (Industry)

9 – Leonard (Len) Salvig, RE Powered Inc. (Industry)

10 – Erik Limpaecher, MIT Lincoln Laboratory, Energy Initiative (Education)

11 – Krista A. Kisch, Vice President – Project Development, BrightSource Energy, Inc. (Industry)

12 – Tom Vinson, AWEA (Industry Trade Group)

13 – Andrea Schiavino, Naval Facilities Engineering Command (Government Agency, user)

14 – Brian Small, Vice President, Competitive Power Ventures, Inc. (Industry)

15 – REC Solar (Industry)